Olives and Olive Related Products - Innovations in Production and Processing

Edited by Vasiliki S. Lagouri

Published in London, United Kingdom

Olives and Olive Related Products - Innovations in Production and Processing
http://dx.doi.org/10.5772/10.5772/intechopen.1004683
Edited by Vasiliki S. Lagouri

Contributors
Ahmed Mohamed Taha, Georgios Efthimiou, Hamdy El-Houssainy Khalifa, Maria Lúcia Pato, Maria Trapali, Perihan Adun, Shirin Tarbiat, Vasiliki S. Lagouri, Şebnem Güler

First published in London, United Kingdom, 2025 by IntechOpen
IntechOpen is the global imprint of INTECHOPEN LIMITED, registered in England and Wales, registration number: 11086078, 167-169 Great Portland Street, London, W1W 5PF, United Kingdom

For EU product safety concerns: IN TECH d.o.o., Prolaz Marije Krucifikse Kozulić 3, 51000 Rijeka, Croatia, info@intechopen.com or visit our website at intechopen.com.

British Library Cataloguing-in-Publication Data
A catalogue record for this book is available from the British Library

Olives and Olive Related Products - Innovations in Production and Processing
Edited by Vasiliki S. Lagouri
p. cm.

This title is part of the Agricultural Sciences Book Series, Volume 17
Topic: Agronomy and Horticulture
Series Editor: W. James Grichar
Topic Editor: Ibrahim Kahramanoglu
Associate Topic Editors: Murat Helvaci and Olga Panfilova

Print ISBN 978-0-85466-033-9
Online ISBN 978-0-85466-032-2
eBook (PDF) ISBN 978-0-85466-088-9
ISSN 3029-052X

If disposing of this product, please recycle the paper responsibly.

IntechOpen Book Series
Agricultural Sciences
Volume 17

Aims and Scope of the Series

The importance of agriculture cannot be overstated. It helps sustain life, as it gives us the food we need to survive and provides opportunities for economic well-being. Agriculture helps people prosper around the world and combines the creativity, imagination, and skill involved in planting crops and raising animals with modern production methods and new technologies. This series includes two main topics: Agronomy and Horticulture, and Animal Farming. This series will help readers better understand the intricacies of production agriculture and provide the new knowledge that is required to be successful. The success of a farmer in modern agriculture requires knowledge of events happening locally as well as globally that impact input decisions and ultimately determine net profit.

Meet the Series Editor

W. James Grichar has been employed with Texas A&M AgriLife Research for over 45 years with an emphasis on research in agronomy, plant pathology, and weed science. He obtained his BS from Texas A&M in 1972 and his Masters of Plant Protection in 1975. He has published 195 journal articles, over 330 research reports and briefs, 11 book chapters, and over 300 abstracts of profession meetings. He also directs research in many crops including corn, grain sorghum, peanuts, and sesame. He has held various positions in different professional societies including the American Peanut Research and Education Society, Southern Weed Science Society, and Texas Plant Protection Conference in addition to being Associate Editor for Peanut Science and Weed Technology. Significant accomplishments have included spearheading efforts to determine the optimum planting time for soybean production along the upper Texas Gulf Coast. These efforts have shown growers that soybean yields can be improved by 10 to 20% by following a late March to early April plant date. He also has been instrumental in developing a herbicide program for peanut production in the south Texas growing region. Through the development and use of herbicides that are effective against major weed problems in the south Texas region, peanut yields have increased by 25 to 30%.

Meet the Volume Editor

Vasiliki Lagouri, BA, MSc, Ph.D., received a BA in chemistry from Aristotle University of Thessaloniki, Greece, where the National Fellowships Foundation awarded her the highest undergraduate honor. She received an MSc in medicinal chemistry from the Department of Pharmacy, National and Kapodistrian University of Athens, and a Ph.D. in food chemistry from the Aristotle University of Thessaloniki. She was awarded postgraduate grants from the National Fellowships Foundation and the Mpodosakis Foundation of Greece. She has more than 25 years of academic and research project management experience at the Department of Food Science and Technology at the International Hellenic University of Thessaloniki, the Department of Chemistry and Pharmacy at the National and Kapodistrian University of Athens, and the National Hellenic Research Foundation. From the academic year 2020–21, she has participated as an adjunct lecturer in the postgraduate program "Biomedical Methods and Technology in Diagnosis" in the Department of Biomedical Sciences at the University of West Attica in Athens. She also holds a position as an adjunct lecturer in the Department of Food Science and Technology at Perrotis College (associated with Cardiff Metropolitan University, UK) in American Farm School of Thessaloniki. She has 21 publications and 24 conference participations in the fields of food chemistry, medicinal chemistry, and natural products to her credit.

Contents

Preface

The Introductory chapter, "Olive Products and Health Beneficial Polyphenols", opens the discussion on olives, olive oil, and its by-products, which are increasingly recognized as an important supply of bioactive compounds and strong antioxidants such as polyphenols. These compounds possess health-beneficial properties, making them potential ingredients for developing nutraceutical products. By using unique technological processing methods, the value of olive mill by-products can be upgraded, improving the environmental ecosystem and leading to a potential economic advantage.

The author of the Perspective chapter "Recent Advancements in the Use of Olive Products for Biotechnological Processes", describes the most recent applications of olive-derived products in biotechnology, agriculture, and food technology. What is the usefulness of the lower-quality oils or waste materials from the olive mill except for high-quality virgin olive oils? This review demonstrates that these low-value products can produce many valuable products, such as biofuel, soil fertilizers, medicines, enzymes, food packaging biopolymers, and even building materials and textile dyes. Furthermore, compounds from olive mill wastewater have been found to stimulate plant growth by affecting their metabolism and have potent antimicrobial properties, helping the biocontrol of hazardous plant pathogens. All these applications of olive products and by-products can help establish a sustainable circular economy by recycling waste and resources.

Oleuropein, discussed in the chapter "The Role of Oleuropein on Cosmeceutical Applications for Enhanced Skin Health", is a phenolic compound derived from olive industry by-products with significant contributions to skin health; it has been described as a useful ingredient in cosmeceutical applications.

The chapter "Olive in Egypt: Cultural Practices for Sustainable Production" analyzes the most recent studies on olive production practices, the main impacts of climate change on olive tree cultivation, and the possible mitigation and adaptation strategies against the potentially negative impacts of climate variability under Egyptian conditions.

Concerning the topic of olive oil and its health effects, the chapter "Olive Oil and Diseases" emphasizes the importance of olive oil in the Mediterranean diet and how the consumption of extra virgin olive oil and table olives offers significant benefits to human health. Extra virgin olive oil, a rich source of "good" fats and phenolic antioxidants, helps reduce the risk of cardiovascular diseases, type 2 diabetes mellitus, cancer, and neurodegenerative diseases.

Olive oil tourism, as described in the chapter "Olive Oil Tourism: Innovative or Traditional Form of Rural Tourism?" is popular in Mediterranean countries where olive oil production has deep cultural and historical roots.

Finally, the chapter "Mediterranean Treasures: Olive Varieties for Table and Olive Oil" highlights the cultural, nutritional, and economic importance of olive varieties cultivated across the Mediterranean region. Olive diversity enriches both traditional and modern cuisines with unique flavours and nutritional value, serving as a cornerstone of the Mediterranean diet and providing numerous health benefits. Research emphasizes the adaptability of olive cultivars to various climatic and environmental stressors, supporting the sustainability of olive production despite the challenges posed by global climate change.

Vasiliki S. Lagouri
Laboratory of Chemistry, Biochemistry, Cosmetic Science,
University of West Attica,
Athens, Greece

Perrotis College,
American Farm School,
Thessaloniki, Greece

Chapter 1

Introductory Chapter: Olive Products and Health Beneficial Polyphenols

Vasiliki S. Lagouri

1. Introduction

The biggest producer of olive oil is the Mediterranean region, especially the nations that border the Mediterranean Sea. The nations that make up the European Union are the world's top producers, consumers, and exporters of olive oil, accounting for over 69% of the global supply [1]. Production and consumption of olive oil are predicted to increase as its health benefits become more widely acknowledged.

Olive oil is extracted by following a few steps: (i) crushing the olives to release the oil, (ii) mixing the paste to improve the production of oil, and (iii) separating the oil from the leftover debris. The latter can be carried out *via* 2-phase centrifugal extraction systems, 3-phase centrifugal systems, or conventional discontinuous press techniques [2–4]. Pressure and three-phase centrifugation systems are two of these methods that produce a lot of Olive Mill Wastewater (OMW), up to 30 million cubic meters annually [5]. Because of its acidic pH and high concentration of organic materials like phenols, this waste is extremely polluting [1, 5]. The significant residues pose an issue from an environmental and economic standpoint. Therefore, novel strategies are required for the detoxification and product valorization of the olive sector.

Olive oil extraction procedures are outlined in **Figure 1**.

2. The composition of olive mill wastewater (OMW) polyphenols

Phenolics can be found in olive fruit or may be generated during the process of producing olive oil. Olive oil includes a diverse range of phenolic compounds, including phenolic acids, alcohols, secoiridoids, and flavonoids (**Figure 2**).

Olive Mill Wastewater (OMW) is also a highly abundant source of phenolic compounds, which can be classified into four main categories: phenyl alcohols, phenolic acids, secoiridoids, and flavonoids. The predominant phenolics found in OMW are tyrosol and hydroxytyrosol, as reported [8–10]. The health advantages of OMW are mostly attributed to the substantial antioxidant capacity of phenolic compounds. They function as chain terminators by providing hydrogen radicals to alkylperoxyl radicals, which are formed during lipid oxidation and the formation of stable derivatives in the process. These characteristics have attracted significant interest from the food and pharmaceutical industries, as they have the potential to be used as preventive substances for chronic and degenerative diseases associated with oxidative stress processes (**Figure 3**) [12–14].

IntechOpen

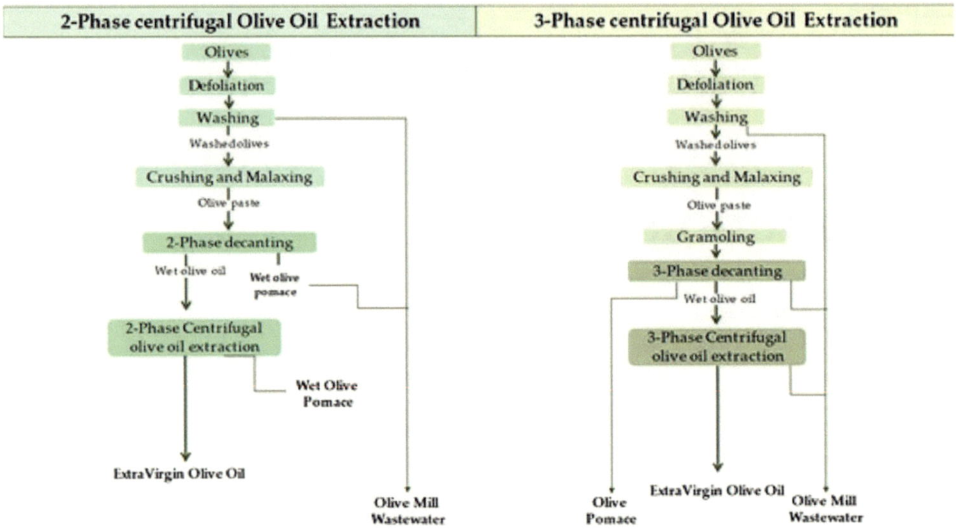

Figure 1.
Olive oil extraction by 2- and 3-phase centrifugation [6].

Figure 2.
A representative scheme evidencing the principal classes of polyphenols in olive oil [7].

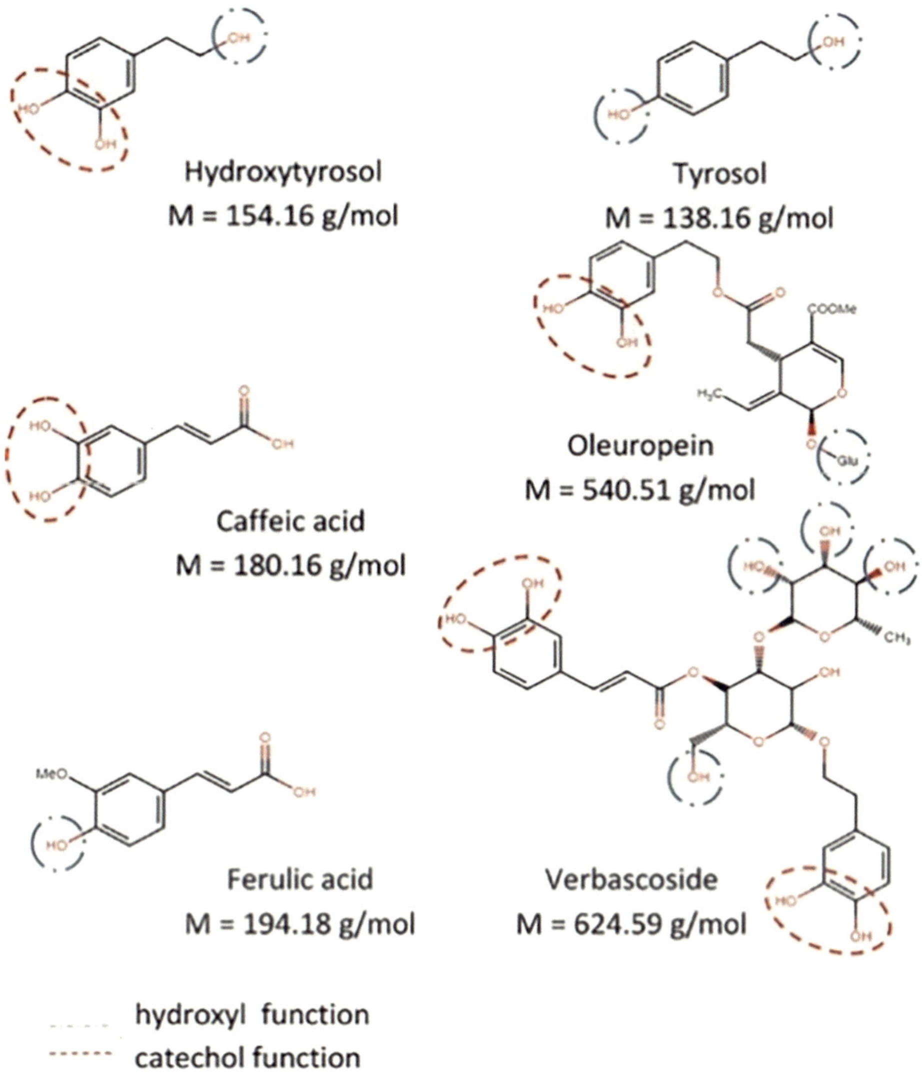

Figure 3.
Primary phenolic compounds in olive mill waste [11].

The olive mill wastewater (OMW) contains around 98% of the phenolics, while only 2% are found in the oil. Recent research has emphasized the potential therapeutic benefits of OMW molecules, based on their biological features, particularly in the field of dermatology [6, 15, 16].

Olive mill wastewater (OMW) contains various phenolic components, including o- and p-coumaric acids, cinnamic acid, caffeic acid, gallic acid, protocatechuic acid, vanillic acid, elenolic acid, tyrosol, and hydroxytyrosol [9, 17, 18]. OMW consists of polyphenols, including oleuropein, demethyloleuropein, and verbascoside [19].

The OMW contains significant quantities of secoiridoid derivatives, including the hydroxytyrosol-bound di-aldehyde of 3,4-dihydroxyphenyl-elenolic acid [20]. The most notable flavonoids that have been recognized are apigenin, cyanidin flavone, anthocyanin, and quercetin [21, 22].

3. Polyphenol bioactivities

Polyphenols, which are antioxidants, combat free radicals, bind to metals, and hinder lipid peroxidation. The activities of DPPH, ABTS, and FRAP assays, as well as biological systems, were demonstrated by Hamden et al. [23] and Carrara et al. [11]. Hydroxytyrosol has been certified by the European Food Safety Authority for its ability to protect against LDL cholesterol and lipid oxidation, as stated by Bulotta et al. [14]. It is advisable to use phenols from olive mill wastewater (OMW), which are rich in antioxidants, in the production of food and cosmetics [24, 25]. Phenols have been investigated for their antioxidant capabilities in living organisms. They can decrease the oxidation of LDL, which is a process that contributes to the development of atherosclerosis. In addition, they reduce DNA oxidation [26, 27].

The main physiologically active compounds in olive mill effluent are o-diphenols, such as hydroxytyrosol, oleuropein, and tyrosol. During laboratory experiments, they demonstrate the ability to decrease LDL oxidation and protect human erythrocytes and DNA from oxidative damage even at low concentrations. Hydroxytyrosol is a nutraceutical that has been acknowledged by the European Food Safety Authority for its ability to protect against LDL cholesterol and lipid oxidation [28]. Hamden et al. [23] demonstrated that in diabetic mice, the reduction of oxidative stress and free radicals, along with the enhancement of enzymatic defenses, resulted in hypoglycemia and antioxidant effects. Hydroxytyrosol exhibits a higher level of antioxidant activity compared to ascorbic acid and BHT [29].

4. OMW sustainability management

Given the environmental difficulties surrounding the disposal of organic municipal waste (OMW), the implementation of sustainable management strategies is crucial. Recently, there has been a suggestion to use the recovery of phenolic compounds as a clever approach to make better use of these by-products, with a focus on circular economy principles.

Current research is exploring techniques such as physicochemical treatment, bioremediation, and bioactive component valorization, as summarized in **Figure 4**. Therefore, the incorporation and execution of these methods could offer a long-lasting resolution to the olive oil industry. The extraction and purification of phenolic compounds from olive mill wastewater (OMW) not only has a positive environmental effect but also yields valuable components for the cosmetics and nutraceutical sectors [6, 30].

5. Conclusion

Olive, olive oil, and its by-products are increasingly recognized as an abundant supply of polyphenols, which offer significant health benefits. The bioactive

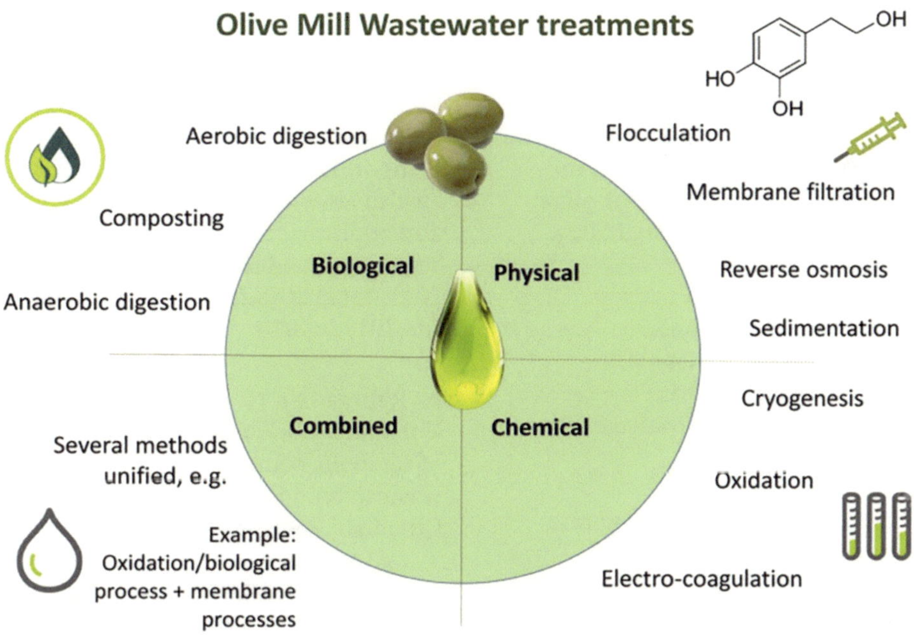

Figure 4.
OMW treatment to lower the pollution charge or extract phenolic compounds for further applications [24].

compounds possess antibacterial, anti-inflammatory, and antioxidant properties, making them valuable for formulating nutraceutical products. Furthermore, by effectively managing and accessing the value of olive mill wastewater (OMW), not only can we enhance human and environmental well-being, but we may also transform an ecological issue into a potential economic advantage.

Author details

Vasiliki S. Lagouri[1,2]

1 Laboratory of Chemistry, Biochemistry, Cosmetic Science, University of West Attica, Athens, Greece

2 American Farm School, Perrotis College, Thessaloniki, Greece

*Address all correspondence to: vlagouri@uniwa.gr

References

[1] Dermeche S, Michaud P, Nadour M, Larroche C, Moulti-Mati F. Biochemical characterizations and techniques for valorization of olive mill wastes. Process Biochemistry. 2013;**48**(10):1532-1552

[2] Roig A, Sanchez-Monedero MA, Cayuela ML. An overview of the techniques used to valorize waste from olive mills. Waste Management. 2006;**26**(9):960-969

[3] Klen T, Vodopivec BM. Phenolic chemicals in olive oil processing: Three-phase decanter versus traditional press and their fates and transformations. LWT-Food Science and Technology. 2012;**49**(2):267-274

[4] El Abbassi A, Zbakh H. Issues with safety and toxicity related to the possible reuse of wastewater from olive mills in agriculture. Waste Management. 2012;**32**(12):2376-2384

[5] El-Abbassi A, Hafidi A, Khayet M. Olive mill effluent can be treated using microporous hydrophobic membranes using direct contact membrane distillation. Desalination. 2012;**287**:9-17

[6] Restuccia D, Prencipe SA, Ruggeri M, Gianfranco Spizzirri U. Sustainability assessment of different extra virgin olive oil extraction methods through a life cycle thinking approach: Challenges and opportunities in the Elaio-technical sector. Sustainability. 2022;**14**:15674. DOI: 10.3390/su142315674

[7] Finicelli M, Squillaro T, Galderisi U, Peluso G. Polyphenols, the healthy brand of olive oil: Insights and perspectives. Nutrients. 2021;**13**:3831. DOI: 10.3390/nu13113831

[8] Araújo M, Pimentel FB, Alves RC, Beatriz M, Oliveira PP. Phenolic compounds from olive mill wastes: Health effects, analytical approach and application as food antioxidants. Trends in Food Science and Technology. 2015;**45**(2):200-211. DOI: 10.1016/j.tifs.2015.06.010

[9] Rahmanian N, Jafari SM, Galanakis CM. Recovery and removal of phenolic compounds from olive mill wastewater. Journal of the American Oil Chemists' Society. 2014;**91**:1-18

[10] Berbel J, Araujo M. A assessment of options for treating and valorizing wastewater from olive mills. Reviews of Renewable and Sustainable Energy. 2018;**79**:70-81

[11] Carrara M, Kelly MT, Roso F, Larroque M, Margout D. Potential of olive oil mill wastewater as a source of polyphenols for the treatment of skin disorders: A review. Journal of Agricultural and Food Chemistry. 2021;**69**(26):7268-7284. DOI: 10.1021/acs.jafc.1c00296

[12] Ambriz-Pérez DL, Leyva-López N, Gutierrez-Grijalva EP, Heredia JB, Yildiz F. Phenolic compounds: Natural alternative in inflammation treatment. A review. Cogent Food and Agriculture. 2016;**2**:1131412. DOI: 10.1080/23311932.2015.1131412

[13] Dias R, Oliveira H, Fernandes I, Simal-Gandara J, Perez-Gregorio R. Recent advances in extracting phenolic compounds from food and their use in disease prevention and as cosmetics. Critical Reviews in Food Science and Nutrition. 2020;**61**(7):1130-1151. DOI: 10.1080/10408398.2020.1754162

[14] Bulotta S, Celano M, Lepore SM, Montalcini T, Pujia A, Russo D. Beneficial effects of the olive oil phenolic components oleuropein and hydroxytyrosol: Focus on protection against cardiovascular and metabolic diseases. Journal of Translational Medicine. 2014;**12**:1-9

[15] Obied HK, Prenzler PD, Konczak I, Rehman AU, Robards K. Chemistry and bioactivity of olive biophenols in some antioxidant and antiproliferative in vitro bioassays. Chemical Research in Toxicology. 2009;**22**:227-234

[16] Schlupp P, Schmidts TM, Pössl A, Wildenhain S, Lo Franco G, Lo Franco A, et al. Effects of a phenol-enriched purified extract from olive mill wastewater on skin cells. Cosmetics. 2019;**6**:30. DOI: 10.3390/cosmetics6020030

[17] Jafari SM, Galanakis CM, Rahmanian N. Recovery and elimination of phenolic chemicals from wastewater produced by olive mills. American Oil Chemists' Society Journal. 2014;**91**(1):1-18

[18] Della Greca M, Previtera L, Temussi F, Zarrelli A. Low molecular-weight components of olive oil mill waste-waters. Phytochemical Analysis. 2004;**15**:184-188

[19] Japon-Lujan R, de Castro MDL. Static-dynamic super heated liquid extraction of hydroxytyrosol and other biophenols from alperujo (a semisolid residue of the olive oil industry). Journal of Agricultural and Food Chemistry. 2007;**55**:3629-3634

[20] Lo Scalzo R, Scarpati ML. A new secoiridoid from olive waste-waters. Journal of Natural Products. 1993;**56**:621-623

[21] Obied HK, Allen MS, Bedgood DR, Prenzler PD, Robards K, Stockmann R. Bioactivity and analysis of biophenols recovered from olive mill waste. Journal of Agricultural and Food Chemistry. 2005;**53**:823-837

[22] De Marco E, Savarese M, Paduano A, Sacchi R. Char acterization and fractionation of phenolic compounds extracted from olive oil mill wastewaters. Food Chemistry. 2007;**104**:858-867

[23] Hamden K, Allouche N, Damak M, Elfeki A. Hypoglycemic and antioxidant effects of phenolic extracts and purified hydroxytyrosol from olive mill waste in vitro and in rats. Chemico-Biological Interactions. 2009;**180**(3):421-432

[24] Caporaso N, Formisano D, Genovese A. Use of phenolic compounds from olive mill wastewater as valuable ingredients for functional foods. Critical Reviews in Food Science and Nutrition. 2018;**58**(16):2829-2841. DOI: 10.1080/10408398.2017.1343797

[25] Foti P et al. Olive mill wastewater as renewable raw materials to generate high added-value ingredients for agro-food industries. Applied Sciences. 2021;**11**(16):7511

[26] Kiokias S, Proestos C, Oreopoulou V. Effect of natural food antioxidants against LDL and DNA oxidative changes. Antioxidants. 2018;7:133. DOI: 10.3390/antiox7100133

[27] Ahmadi A, Jamialahmadi T, Sahebkar A. Polyphenols and atherosclerosis: A critical review of clinical effects on LDL oxidation. Pharmacological Research. 2022;**184**:106414

[28] Silva AFR et al. Application of hydroxytyrosol in the functional foods field: From ingredient to

dietary supplements. Antioxidants. 2020;**9**(12):1246

[29] Aruoma OI et al. Effect of hydroxytyrosol found in extra virgin olive oil on oxidative DNA damage and on low-density lipoprotein oxidation. Journal of Agricultural and Food Chemistry. 1998;**46**(12):5181-5187

[30] Gómez-Caravaca AM et al. Recovery of phenolic compounds from olive oil mill wastewaters by physicochemical methodologies. In: Grumezescu AM, Holban AM, editors. Handbook of Food Bioengineering, Food Bioconversion. Academic Press; 2017. pp. 467-489. ISBN 9780128114131

Chapter 2

Perspective Chapter: Recent Advancements in the Use of Olive Products for Biotechnological Processes

Georgios Efthimiou

Abstract

This chapter discusses the most recent applications of olive-derived materials in biotechnology, agriculture and food technology. Although the high-quality olive oil is usually destined for human consumption, lower oil grades or waste materials from the olive mill can be used to produce a plethora of valuable products, such as biofuel, soil fertilisers, medicines, enzymes, food packaging biopolymers and even building materials and textile dyes. Furthermore, compounds from olive mill wastewater have been found to stimulate plant growth by affecting their metabolism and also have potent antimicrobial properties, helping the biocontrol of hazardous plant pathogens. All these applications of olive products and by-products can help establish a sustainable circular economy through recycling of waste and resources.

Keywords: olive, olive mill wastewater, olive oil, biotechnology, high-value products, waste valorisation, circular economy

1. Introduction

1.1 Olive products and by-products

The olive is the fruit of the olive tree (*Olea europaea* L.), which is mainly found in the Mediterranean basin, but also cultivated in other areas of the world that allow its growth, such as North and South America, South Africa, Australia and New Zealand [1]. Traditionally, the flesh of the fruit is used for the production of olive oil, which is destined for human consumption in a variety of dietary applications [1].

The kernel of the olive fruit is often used to produce olive kernel (or pomace) oil, an oil which is considered of lower dietary quality compared to olive oil [2].

The waste produced during the olive oil extraction procedure is called olive mill wastewater (OMW) and has been identified as useful material for the synthesis of various key products *via* several pathways of waste valorisation [3].

1.2 Vegetable oils in bio-fermentation

Olive oil, as well as other vegetable oils has a long history of use in microbial fermentation in large industrial bioreactors, where the oil is slowly added as a carbon source during the fermentation process [4, 5]. This can lead to the production of many high-value products such as antibiotics, organic acids or enzymes [6–8].

1.3 Pyrolysis of OMW

OMW can be treated with pyrolysis, a process that can be used to convert olive mill wastewater (OMW) into bio-oil, biochar and syngas [9]. Different pyrolysis methods have been recently suggested, such as fast pyrolysis and low-temperature microwave-assisted pyrolysis [10, 11]. These processes have been supported with modelling approaches in order to predict and optimise product yield [12]. Pyrolysis products from OMW are traditionally used as biofuel or soil amenders for agricultural purposes.

1.4 Anaerobic digestion for biogas production

Anaerobic digestion of OMW can also lead to useful products, such as methane [13]. During anaerobic digestion, the OMW sludge is being digested by microbes that eventually produce different types of biogas under anaerobic conditions. Methane and other useful gases can be then utilised for production of energy and/or electricity.

1.5 Range of products synthesised from olive-derived materials

In this chapter, new developments in biotechnological applications of olive products and by-products will be highlighted and discussed. In Section 2, different types of valuable products from olive-derived starting materials will be portrayed, including biosurfactants and biolubricants (Section 2.1), high-value products such as antibiotics, enzymes and organic acids (Section 2.2), agricultural (Section 2.3) and food technology (Section 2.4) applications and finishing with novel innovative products such as leather dyes, building materials and polyurethane foam (Section 2.5).

2. Recent advancements in the use of olive products in biotechnology

2.1 Bio-lubricants and biosurfactants

Olive oil itself has lubricating and surfactant (emulsifying) properties and can be used in a variety of industrial and food-related applications. It also prevents formation of foam in bioreactors with microbial cultures [6]. However, due to its high price, other olive by-products are often used to synthesise bio-lubricants and biosurfactants. Lubricant dry substances or coating is used to reduce friction between two surfaces, while surfactants can be used in several homogenisation processes in the food industry or for facilitating digestion of petroleum spills by microbes [14, 15]. Olive-derived materials have been recently reported to lead to the synthesis of such agents [16, 17].

2.2 High-value products

Olive products and by-products have been described to be able to lead to the biosynthesis of high-value compounds when used as carbon sources in biofermentation vessels with microbial cultures [18]. In these cases, the microbes (usually bacteria, yeast or filamentous fungi) utilise the vegetable oil or waste for supporting their metabolism, increasing their biomass and synthesising key primary or secondary products. For example, the antibiotic clavulanic acid is known to be produced in high yields when olive or olive kernel oil is added in the culture [6, 19]. Enzymes, such as lipases, can be efficiently produced by microbes in oil-containing media and subsequently be extracted and purified [20]. The yeast *Yarrowia lipolytica* has been recently reported to facilitate the biosynthesis of ethanol (biofuel, beverages and research), citric acid (important for the food industry) and magnesium oxide nanoparticles (antibacterial activity) from olive-derived materials [21–24]. Lastly, utilisation of exhausted olive pomace by *Crypthecodinium cohnii* led to the synthesis of omega-3 fatty acids [25], while oleochemical modification by application of ascorbic acid led to the synthesis of bio-based cyclic carbonates that can be used as green biofuels [26].

2.3 Agricultural applications

Olive-derived materials have also been used in agriculture as fertilisers and biocontrol agents. For instance, OMW can be used as fertiliser for crops, while it can also be treated for production of clean water that can be used for irrigation of crops [27]. Olive waste composts have also been applied for high-yield growth of tomato seedlings [28]. Other groups have previously tested similar composts from olive pomace or leaves on lettuce, tomato and emmer seedlings, all reporting higher yields compared to the untreated control crops [29–31]. Finally, OMW is a source of bioactive molecules for plant growth and protection against pathogens [32]. Humic acids extracted from OMW can be used as plant biostimulant in agriculture, thanks to their positive effects on plant biomass production, nutrition and activity of enzymes implied in N metabolism and glycolysis. Drobek et al. [33] showed that plant biostimulant formed as a by-product of the olive oil can induce an increase in the protein content of maize grains up to 19%.

Phenols from OMW have similar biostimulating effects on plants [33]. Finally, OMW water has been found to have strong antifungal properties that protect crops against *Fusarium oxysporum*, *Pythium* spp., *Sclerotinia sclerotiorum*, *Verticillium dahlia* and *Botrytis cinerea* [34, 35].

2.4 Food technology applications

Valorization of by-products from olive oil industry has also led to the development of a variety of innovative functional foods [36, 37]. Olive oil itself has been used together with potato starch for creating edible films reinforced with nanoparticles for their use as edible food packaging [38]. Moreover, phenols from olive oil by-products have been employed for food applications through microencapsulation approaches [39]. More specifically, most of phenolic compounds present in olive oil by-products lose their activity quite quickly due to environmental factors such as temperature, pH and light exposure. Therefore, prior to food fortification with phenolic-rich extracts obtained from olive oil by-products, they should be protected through microencapsulation approaches, allowing a sustained release of phenolic compounds in the fortified foods. Finally, more examples of food packaging from olive pomace-based biocomposites and other alternatives to conventional petrochemical-based packaging materials often reinforced with plant essential oils, natural additives, improved oxygen barrier, and antibacterial and antifungal properties [40, 41].

2.5 Innovative manufacturing applications

A variety of novel products based on olive-derived materials have been developed recently. These include the use of olive oil polyols for making renewable polyurethane foams [42] and bio-based insulating building materials from recycled olive core [43]. Finally, leather and textile dyes can be developed from OMW that are of great importance for the tanning industry [44].

3. Conclusions

In this chapter, several examples of biotechnological applications of olive products and by-products were presented and discussed. It is clear that these materials can be exploited *via* different pathways and lead to the synthesis of a wide variety of valuable products. Several parts of the olive tree such as the wood and the leaves can be used for energy generation, while the olive fruit and its core can also be used for making medicines, fertilisers, biofuel, food coatings and even building materials and textile dyes. Further research will definitely suggest more such applications in the near future. All these examples make olive a precious fruit for the biotech industry and a significant player in the maintenance of a sustainable circular economy, aiming to reduce waste and conserve resources by recycling and reusing products and components [45].

Author details

Georgios Efthimiou
Centre for Biomedicine, Hull York Medical School, Hull, UK

*Address all correspondence to: g.efthimiou@hull.ac.uk

References

[1] Soares TF, Alves RC, Oliveira MBPP. From olive oil production to by-products: Emergent technologies to extract bioactive compounds. Food Reviews International. 2024:1-28. Available from: https://www.tandfonline.com/doi/full/10.1080/87559129.2024.2354331

[2] Roy S, Sarkar T, Upadhye VJ, Chakraborty R. Comprehensive review on fruit seeds: Nutritional, phytochemical, nanotechnology, toxicity, food biochemistry, and biotechnology perspective. Applied Biochemistry and Biotechnology. 2024;**196**(7):4472-4643

[3] Castro-Muñoz R, Díaz-Montes E, Gontarek-Castro E, Boczkaj G, Galanakis CM. A comprehensive review on current and emerging technologies toward the valorization of bio-based wastes and by products from foods. Comprehensive Reviews in Food Science and Food Safety. 2022;**21**(1):46-105

[4] Cavanagh ME, Ison AP, Lilly MD, Carley-Smith SW, Edwards J. The utilization of lipids during Streptomyces fermentations. Transactions. Institute of Chemical Engineers. 1994;**72**:14-16

[5] Large KP, Ison AP, Williams DJ. The effect of agitation rate on lipid utilisation and clavulanic acid production in *Streptomyces clavuligerus*. Journal of Biotechnology. 1998;**63**:111-119

[6] Efthimiou G, Thumser AE, Avignone-Rossa CA. A novel finding that *Streptomyces clavuligerus* can produce the antibiotic clavulanic acid using olive oil as a sole carbon source. Journal of Applied Microbiology. 2008;**105**(6):2058-2064

[7] Barrington S, Kim J-W. Response surface optimization of medium components for citric acid production by *Aspergillus niger* NRRL 567 grown in peat moss. Bioresource Technology. 2008;**99**(2):368-377

[8] Gombert AK, Pinto AL, Castilho LR, Freire DMG. Lipase production by *Penicillium restrictum* in solid-state fermentation using babassu oil cake as substrate. Process Biochemistry. 1999;**35**(1-2):85-90

[9] Rajpoot L, Tagade A, Deshpande G, Verma K, Geed SR, Patle DS, et al. An overview of pyrolysis of de-oiled cakes for the production of biochar, bio-oil, and pyro-gas: Current status, challenges, and future perspective. Bioresource Technology Reports. 2022;**19**:1-14

[10] Dorado F, Sanchez P, Alcazar-Ruiz A, Sanchez-Silva L. Fast pyrolysis as an alternative to the valorization of olive mill wastes. Journal of the Science of Food and Agriculture. 2021;**101**(7):2650-2658

[11] Bartoli M, Rosi L, Giovannelli A, Frediani P, Frediani M. Characterization of bio-oil and bio-char produced by low-temperature microwave-assisted pyrolysis of olive pruning residue using various absorbers. Waste Management & Research. 2020;**38**(2):213-225

[12] Makepa DC, Chihobo CH, Musademba D. Techno-economic analysis and environmental impact assessment of biodiesel production from bio-oil derived from microwave-assisted pyrolysis of pine sawdust. Heliyon. 2023;**9**(11):e22261

[13] Sampaio MA, Gonçalves MR, Marques IP. Anaerobic digestion challenge of raw olive mill wastewater. Bioresource Technology. 2011;**102**(23):10810-10818

[14] Uppar R, Dinesha P, Kumar S. A critical review on vegetable oil-based bio-lubricants: Preparation, characterization, and challenges. Environment, Development and Sustainability. 2023;**25**:9011-9046

[15] Sarubbo LA, Silva MC, Durval IJB, Bezerra KGO, Ribeiro BG, Silva IA, et al. Biosurfactants: Production, properties, applications, trends, and general perspectives. Biochemical Engineering Journal. 2022;**181**:1-19

[16] Masripan NAB, Salim MA, Omar G, Mansor MR, Saad A, Hamid NA, et al. Vegetable oil as bio-lubricant and natural additive in lubrication: A review. International Journal Of Nanoelectronics And Materials. 2020;**13**:161-176

[17] Chebbi A, Franzetti A, Duarte Castro F, et al. Potentials of winery and olive oil residues for the production of rhamnolipids and other biosurfactants: A step towards achieving a circular economy model. Waste and Biomass Valorization. 2021;**12**:4733-4743

[18] del Mar Contreras M, Romero I, Moya M, Castro E. Olive-derived biomass as a renewable source of value-added products. Process Biochemistry. 2020;**97**:43-56

[19] Young T, Li Y, Efthimiou G. Olive pomace oil can be used as an alternative carbon source for clavulanic acid production by *Streptomyces clavuligerus*. Waste and Biomass Valorization. 2020;**11**:3965-3970

[20] Adetunji AI, Olaniran AO. Production strategies and biotechnological relevance of microbial lipases: A review. Brazilian Journal of Microbiology. 2021;**52**(3):1257-1269

[21] Sarris D, Rapti A, Papafotis N, Koutinas AA, Papanikolaou S. Production of added-value chemical compounds through bioconversions of olive-mill wastewaters blended with crude glycerol by a *Yarrowia lipolytica* strain. Molecules. 2019;**24**(2):222

[22] Azaizeh H, Tayeh HNA, Gerchman Y. Chapter 2—Valorisation of olive oil industry solid waste and production of ethanol and high value-added biomolecules. In: Rathinam NK, Sani RK, editors. Biovalorisation of Wastes to Renewable Chemicals and Biofuels. Elsevier; 2020. pp. 27-40

[23] PapanikolaouS,Galiotou-PanayotouM, Fakas S, Komaitis M, Aggelis G. Citric acid production by *Yarrowia lipolytica* cultivated on olive-mill wastewater-based media. Bioresource Technology. 2008;**99**(7):2419-2428

[24] Hamimed S, Jebli N, Sellami H, Landoulsi A, Chatti A. Dual valorization of olive mill wastewater by bio-nanosynthesis of magnesium oxide and *Yarrowia lipolytica* biomass production. Chemistry & Biodiversity. 2020;**17**(3):e1900608

[25] Paz A, Karnaouri A, Templis CC, Papayannakos N, Topakas E. Valorization of exhausted olive pomace for the production of omega-3 fatty acids by *Crypthecodinium cohnii*. Waste Management. 2020;**118**:435-444

[26] Natongchai W, Pornpraprom S, D'Elia V. Synthesis of bio-based cyclic carbonates using a bio-based hydrogen bond donor: Application of ascorbic acid to the cycloaddition of CO_2 to oleochemicals Asian. The Journal of Organic Chemistry. 2020;**9**:801

[27] Jeguirim M, Goddard M-L, Tamosiunas A, Berrich-Betouche E, Azzaz AA, Praspaliauskas M, et al. Olive mill wastewater: From a pollutant to green fuels, agricultural water source

and bio-fertilizer. Biofuel production. Renewable Energy. 2020;**149**:716-724

[28] Tüzel Y, Ekinci K, Öztekin GB, Erdal İ, Varol N, Merken Ö. Utilization of olive oil processing waste composts in organic tomato seedling production. Agronomy. 2020;**10**:797

[29] Michailides M, Christou G, Akratos CS, Tekerlekopoulou AG, Vayenas DV. Composting of olive leaves and pomace from a three-phase olive mil plant. International Biodeterioration and Biodegradation. 2011;**65**:560-564

[30] Killi D, Kavdır Y. Effects of olive solid waste and olive solid waste compost application on soil properties and growth of *Solanum lycopersicum*. International Biodeterioration and Biodegradation. 2013;**82**:157-165

[31] Diacono M, Montemurro F. Olive pomace compost in organic emmer crop: Yield, soil properties, and heavy metals' fate in plant and soil. Journal of Soil Science and Plant Nutrition. 2019;**19**:63-70

[32] Sciubba F, Chronopoulou L, Pizzichini D, Lionetti V, Fontana C, Aromolo R, et al. Olive mill wastes: A source of bioactive molecules for plant growth and protection against pathogens. Biology. 2020;**9**:450

[33] Drobek M, Frąc M, Cybulska J. Plant biostimulants: Importance of the quality and yield of horticultural crops and the improvement of plant tolerance to abiotic stress—A review. Agronomy. 2019;**9**:335

[34] Vagelas I, Kalorizou H, Papachatzis A, Botu M. Bioactivity of olive oil mill wastewater against plant pathogens and post-harvest diseases. Biotechnology and Biotechnological Equipment. 2009;**23**:1217-1219

[35] Vagelas I, Papachatzis A, Kalorizou H, Wogiatzi E. Biological control of *Botrytis* fruit rot (*Gray Mold*) on strawberry and red pepper fruits by olive oil mill wastewater. Biotechnology and Biotechnological Equipment. 2009;**23**:1489-1491

[36] Gullón P, Gullón B, Astray G, Carpena M, Fraga-Corral M, Prieto MA, et al. Valorization of by-products from olive oil industry and added-value applications for innovative functional foods. Food Research International. 2020;**137**:109683

[37] Foti P, Romeo FV, Russo N, Pino A, Vaccalluzzo A, Caggia C, et al. Olive mill wastewater as renewable raw materials to generate high added-value ingredients for agro-food industries. Applied Sciences. 2021;**11**:7511

[38] Farajpour R, Emam Djomeh Z, Moeini S, Tavakolipour H, Safayan S. Structural and physico-mechanical properties of potato starch-olive oil edible films reinforced with zein nanoparticles. International Journal of Biological Macromolecules. 2020;**149**:941-950

[39] Paulo F, Santos L. Deriving valorization of phenolic compounds from olive oil by-products for food applications through microencapsulation approaches: A comprehensive review. Critical Reviews in Food Science and Nutrition. 2021;**61**(6):920-945

[40] Banat R, Aljnaid M, Al-Rawashdeh M. Mechanical, morphological and water absorption properties of polyethylene/olive pomace flour bio-composites. Polyolefins Journal. 2022;**9**(1):1-14. DOI: 10.22063/poj.2021.2903.1184

[41] Asgher M, Qamar SA, Bilal M, Iqbal HMN. Bio-based active food

packaging materials: Sustainable alternative to conventional petrochemical-based packaging materials. Food Research International. 2020;**137**:109625

[42] Elena C, Peyrton J, Hubca G, Sarbu A, Gabor R, Nicolae C-A, et al. Synthesis and characterization of renewable polyurethane foams using different biobased polyols from olive oil. European Polymer Journal. 2021;**149**:1-10

[43] Boukhelkhal D, Guendouz M, Bourdot A, et al. Elaboration of bio-based building materials made from recycled olive core. MRS Energy & Sustainability. 2021;**8**:98-109

[44] Meksi N, Haddar W, Hammami S, Mhenni MF. Olive mill wastewater: A potential source of natural dyes for textile dyeing. Industrial Crops and Products. 2012;**40**:103-109

[45] Donner M, Erraach Y, López-i-Gelats F, Manuel-i-Martin J, Yatribi T, Radić I, et al. Circular bioeconomy for olive oil waste and by-product valorisation: Actors' strategies and conditions in the Mediterranean area. Journal of Environmental Management. 2022;**321**:1-12

Chapter 3

The Role of Oleuropein on Cosmeceutical Applications for Enhanced Skin Health

Shirin Tarbiat

Abstract

The biggest organ in the human body, the skin, is essential for defense against physiological and environmental threats. The skin barrier plays a crucial role in keeping the skin healthy and in delivering substances applied topically to the deeper layers of the skin. Cosmeceuticals provide effective skin care treatments, a hybrid product category that falls between cosmetics and pharmaceuticals. Using olive leaves and oils is an inventive way to use goods from the olive industry to cosmetics. This chapter explores an innovative approach that applies olive industry byproducts, like olive oils and leaves, to cosmetic uses. The focus is on oleuropein, a phenolic compound derived from these sources, demonstrating significant potential in skin health. Oleuropein has been shown in numerous animal and clinical trials to have various biological benefits, such as antioxidant, anti-inflammatory, anti-aging, anti-collagenase, anti-elastase, anti-tyrosinase, and anti-hyaluronidase properties. These characteristics help to improve wrinkles, decrease pigmentation, enhance elasticity, decrease skin thickness, and speed up the healing of wounds. In conclusion, studies on oleuropein's function in skin health highlight this compound's potential as a useful ingredient for cosmeceutical products that aim to prevent skin aging and enhance general skin health.

Keywords: oleuropein, skin health, cosmeceutical, antioxidant, skin aging

1. Introduction

The skin, the body's largest organ, serves as the primary defense against environmental factors. It fulfills numerous vital functions, including protection, insulation, regulation of water balance, sensory perception, and vitamin D production. Among these, its foremost role is safeguarding the body from various external threats. Healthy skin is characterized by a smooth, evenly pigmented surface free of irritations, with an efficient pH balance and the ability to quickly repair minor injuries. Additionally, it exhibits elasticity, which helps prevent sagging and wrinkling [1].

Various skin conditions include sunburn, hyperpigmentation, premature aging, dryness, eczema, rosacea, and acne can be caused by various external and internal factors such as pollutants, UV radiation, the biologic progression of cells, nutritional deficiencies, and hormonal imbalances. These issues can persist throughout life, and while some can be mitigated with cosmeceuticals and nutraceuticals, others may require medicinal treatments [2].

Various mechanisms can influence skin problems such as formation of free radicals and inflammation, over activation of related enzymes such as metalloproteases, tyrosinase, collagenase, and elastase play a crucial role in skin pigmentation, degradation of skin cells as well as collagen and elastic fiber breakdown [3].

Cosmeceuticals are a specialized category of skincare products that function as active cosmetics to enhance overall skin health. These topically applied products contain biologically active ingredients that lie on the spectrum between cosmetics and drugs. They are designed to improve skin tone, texture, and radiance, reduce the appearance of wrinkles, and offer anti-aging benefits [4].

Oleuropein, primarily found in olive leaves, belongs to the secoiridoid subclass of phenolic compounds. It is synthesized via the secondary metabolism of terpenes and is composed of three structural subunits: hydroxytyrosol, elenolic acid, and a glucose molecule [5].

Research has shown that oleuropein has numerous biological functions, including antioxidant, antibacterial, anti-inflammatory, anticancer, antidiabetic, cardioprotective, and hepatoprotective effects. It is also effective in treating neurodegenerative diseases such as Alzheimer's and Parkinson's diseases. Various in vitro, in vivo, and clinical studies have confirmed the health benefits of oleuropein, making it a significant subject of research in the fields of pharmaceuticals, food, and cosmetics [6].

In response to consumer demand, the cosmetic industry has increasingly sought natural, highly effective active ingredients for cosmeceutical [7]. This chapter explores how oleuropein, an active ingredient derived from olive leaves and oil, could be effective ingredient to improve skin texture, enhance radiance, reduce acne, diminish the appearance of wrinkles, and provide anti-aging benefits within cosmeceutical formulations.

2. The skin

The skin performs many vital functions, protecting inner organs, maintaining homeostasis, and serving as the primary defense against environmental factors such as physical, chemical, and biological attackers. Thermoregulation, maintaining water balance, sensory regulation, and vitamin D production are other important role of skin. However, unhealthy skin conditions can arise due to a multitude of factors, including genetic and environmental exposures, aging, metabolic processes, and lifestyle choices. These factors significantly change the skin structure and functioning. Common causative agents include excessive UV radiation, pollutants, and chemicals. Additionally, poor nutrients, stress, and inadequate skincare routines can compromise skin integrity. Understanding these factors is essential for maintaining healthy skin and delaying aging. Effective skin health strategies emphasize healthy habits and early intervention [1].

Skin health can be maintained by understanding and targeting the function of the skin barrier and the different layers of the skin.

2.1 Skin layers

The human skin barrier is an important part of the skin's intactness and functions in the transportation of topically applied compounds into deeper skin layers. The skin is composed of three layers: the epidermis, the dermis, and subcutaneous tissue. The epidermis, which is the outermost layer of the skin possesses a barrier function that eliminate bacteria toxins and ultraviolet light in healthy skin. The physical appearance of skin such as color, softness, and dryness are reflected by the epidermis. The stratum corneum is the outermost layer of the epidermis. Consisting of dead tissue, it has a role in hydration of skin as well as the protection of underlying tissue from infection and mechanical stress. It involves the delivery of active compounds of cosmeceuticals across the skin barrier. The epidermis consists of three cell types: keratinocytes, melanocytes, and Langerhans cells. The middle layer, the dermis, is fundamentally composed of mucopolysaccharide gel held together by a fibrous matrix of proteins known as collagen and elastin, which are responsible for the skin's strength and elasticity. The cell types are present in dermis including mast cells, macrophages, and fibroblasts. Collagens are produced by fibroblasts covers 70% of the dry weight of dermis. The dermis is also composed of extracellular matrix, which is made up of insoluble collagen and elastin. Glycosaminoglycans such as hyaluronic acid and dermatan sulfate are macromolecules that are important for skin hydration. The dermis lies on the subcutaneous tissue, which contains lipocytes, giving the skin a lubricant and attractive appearance [8]. The structure and function of the skin layers are crucial to comprehending the complex process of skin aging, which involves changes at multiple levels of the skin.

2.2 Skin aging

Skin aging is a complex biological process influenced by intrinsic (biological aging) and extrinsic (photoaging, environmental toxins, and infectious agents) aging. There is distinct difference between biological aging and photoaging. Biological modification of the skin is characterized by fine lines and wrinkling, loss of elasticity, and rough-textured appearance. In contrast, photoaged skin is associated with uneven pigmentation and is markedly wrinkled. In biological aging, skin functions slowly. The thickness of epidermis decreases, and there is a decline in the number of melanocytes. General atrophy of extracellular matrix occurs in skin tissue, due to decrease in number of fibroblasts along with decrease in collagen and elastin levels and increase in extracellular protein breakdown in dermis, which leads to changes in strength of aging skin. In contrast to biologically aged skin, in photoaging condition, hypertrophy in skin tissue occurs due to increase in activity of certain enzymes in dermis. Hyperplasia of elastic tissue led to accumulation of degraded elastic fibers. The degree of elastosis correlates with the amount of sun exposure. Along with hyperplasia of elastic tissue, glycogen degradation, and glycosaminoglycan level increase in dermis. Collagen degraded due to an increase in the activity of collagenase enzymes in response to UV radiation. Hyperplasia of melanocytes in epidermis causes abnormal skin pigmentation in photodamaged skin. Oxidative stress is considered as a primary cause of both types of skin aging processes. In intrinsic aging, reactive oxygen species (ROS) are produced mainly through cellular oxidative metabolism during adenosine triphosphate (ATP) generation from glucose and lipid metabolism, whereas in extrinsic aging, loss of redox equilibrium is caused by environmental factors, such as ultraviolet radiation, pollution and cigarette smoking [9].

Despite the effects of aging, it is possible to maintain healthy skin throughout life by selecting a proper lifestyle and using nutraceuticals and cosmeceuticals. Skin maintenance must be based on biological properties of the skin therefore it must be treated with agents that target different layers and cells of the skin.

3. Cosmeceuticals

Cosmeceuticals are a hybrid category of products lying on the spectrum between pharmaceuticals and cosmetics. It is now commonly used to describe a cosmetic product that exerts a pharmaceutical therapeutic benefit but not necessarily a biological therapeutic benefit. The difference between a drug and a cosmeceutical is that the former is defined as having a biological effect on living tissue. Cosmeceuticals contain active ingredients like vitamins, antioxidants, or botanical extracts. These ingredients are chosen to enhance skincare and make noticeable changes to the skin.

In recent years, the growth of the cosmeceutical market has increased. Consumers are often interested in cosmeceuticals possessing plant-based active ingredients. Indeed, it is often difficult to separate the effects of the moisturizer vehicle from the effects of the added active ingredient in cosmeceuticals. Understanding the mechanisms of action of cosmeceuticals is important. Cosmeceuticals can be categorized based on the type of active compounds: antioxidants, peptides, anti-inflammatories, polysaccharides, and lightening agents [10].

Free radicals are formed frequently in skin tissue due to metabolisms of biomolecules and mitochondrial dysfunction. They are highly reactive species with unpaired electrons. They target the cell membranes as well as cellular lipids, proteins, and DNA that eventually damage extracellular matrix materials, especially collagen fibers. Antioxidants with various potential and different skin penetration abilities includes vitamins such as vitamins A, E, C, and B group vitamins, vitamin-like compounds, for instance, alpha lipoic acid, coenzyme Q-10, and plant secondary metabolites known as phytochemicals including quercetin, apigenin, curcumin, epigallocatechin are frequently use as an active component in cosmeceuticals, they aim to protect and treat the skin from photodamage, pigmentation, formation of wrinkles. They have a role in the inhibition of metalloprotease enzymes and collagen breakdown [11].

The other active ingredients of cosmeceuticals are growth factors, which are regulatory proteins involved in wound healing such as recombinant epidermal growth factor, placental and cultured fibroblasts derived growth factors function in new collagen, elastin, and glycosaminoglycan production. Cosmeceuticals also contain peptides as a functional ingredient. Peptides are a short chain of (2–50) amino acids joined by covalent bonds. They are components of large proteins like collagen. They can penetrate the skin layers, help form skin proteins in the dermis, and enhance wound healing. Organic acids include lactic acid, malic acid, citric acid, and salicylic acid and are also considered as a functional agent in cosmeceuticals. They can hydrate the skin by increasing glucosamine glycans as well as exfoliate the skin and improve the stratum corneum barrier [12].

4. Oleuropein in skin health

4.1 Oleuropein

Olive (*Olea europaea* Linn., family Oleaceae), commonly known as Zeytoon in the Mediterranean region, is a cornerstone of the Mediterranean diet, primarily due to its fruits and oil. Nutritionally, olives are rich in fats, particularly monounsaturated fats like oleic acid, as well as polyunsaturated and saturated fats. They are also excellent sources of vitamins and minerals, including vitamins E, A, K, iron, calcium, and magnesium. Moreover, olives are abundant in secondary metabolites from various phytochemical classes such as phenolic acids, flavonoids, secoiridoids, and terpenoids. The predominant phytochemicals in olives are phenolic compounds, with oleuropein, hydroxytyrosol, and tyrosol being the most significant for their health benefits [5].

Oleuropein, primarily found in olive leaves, belongs to the secoiridoid subclass of phenolic compounds. It is synthesized via the secondary metabolism of terpenes and is composed of three structural subunits: hydroxytyrosol, elenolic acid, and a glucose molecule. In this structure, elenolic acid is ester-bonded to hydroxytyrosol and glycosidically bonded to glucose. Hydrolysis of these bonds releases the individual molecules. Oleuropein is sensitive to heat, light, and oxygen, becoming unstable at temperatures above 80°C. It is crystalline, odorless, relatively water-soluble, and has low oil solubility, with a chemically large structure that imparts a bitter taste. Besides olives, oleuropein is also found in plants like *Fraxinus excelsior*, *Ligustrum ovalifolium*, *Jasminum polyanthum*, *Syringa josikaea*, and *S. vulgaris* [13].

The oleuropein content in olive fruits and leaves varies depending on cultivar types, environmental factors, and developmental stages. Numerous extraction methods have been developed to maximize the yield of oleuropein and its metabolites from olive-derived materials, including organic solvent extraction, aqueous solvent extraction, ultrasound-assisted extraction (UAE), enzyme-assisted extraction (EAE), solid-phase extraction (SPE), pressurized liquid extraction (PLE), supercritical fluid chromatography (SFC), high-performance liquid chromatography (HPLC), and microwave irradiation techniques [14].

4.2 Oleuropein and the cosmetic industry

In recent years, the cosmetic industry has progressively focused on the application of sustainable and environmentally friendly raw materials. A key innovative approach involves utilizing byproducts from food waste and transforming them into valuable ingredients for cosmeceuticals. This approach not only addresses the issue of resource depletion but also helps reduce food waste accumulation, which poses significant socio-economic and environmental challenges.

One of the major byproducts of the olive oil industry is olive leaves. These leaves generate large amounts of waste, which can contribute to soil and water pollution. To minimize the environmental impact, various strategies for reappraising these byproducts have been proposed. In this context, developing oleuropein-enriched extracts from olive byproducts presents a promising opportunity. Oleuropein, a bioactive compound, can serve as a potential source of ingredients for cosmetic use, providing an economic advantage and adding high cosmetic value [15].

The field of skincare products and cosmetics stands to benefit particularly from these remaining materials. Oleuropein's bioactive compounds can fulfill essential cosmetic functions and activities, making them suitable for use in formulations aimed at improving skin health and appearance. The health benefits of oleuropein, including its antioxidant, anti-inflammatory, and anticancer properties, are well documented. These properties contribute to its potential effectiveness in anti-aging, skin protection, and rejuvenation applications [16].

Given these advantages and the supporting evidence from in vitro, in vivo, and clinical studies, we aimed to explore oleuropein's contribution as an active ingredient in the cosmetic industry. By harnessing the power of this natural compound, the industry can move toward more sustainable practices while offering consumers products that deliver real benefits. As consumer demand for natural and effective skincare solutions continues to grow, oleuropein-enriched formulations could play a pivotal role in the next generation of cosmetic products.

4.3 Safety and cytotoxicity of oleuropein in cosmetic formulations

Assessing the safety and stability of active ingredients is crucial for their application in cosmeceuticals. This ensures that formulations are both effective and free from adverse effects. Oleuropein, a key compound derived from olive oil industry byproducts, has been the subject of extensive research to evaluate its safety profile in cosmetic formulations.

In a study conducted by Andreia Nunes et al., toxicity assays and pH stability tests were performed on extracts derived from olive oil industry byproducts to ensure the absence of irritant conditions. The researchers first assessed the in vitro cytotoxicity, inhibition of skin enzymes, and antioxidant and photoprotective capacities of the extracts. They then formulated oil-in-water creams containing three different olive oil industry byproduct extracts and examined their compatibility, acceptability, and antioxidant efficacy [17]. The results indicated that the cream with the highest concentrations of phenolic compounds exhibited the greatest antioxidant efficiency without any cytotoxic properties. Furthermore, no adverse reactions were observed following the application of these formulations on the skin. The study concluded that these extracts were successfully incorporated into creams, demonstrating favorable appearance, pH stability, and rheological performance.

In another study focusing on cell viability, olive leaf extract enriched with oleuropein was examined for cytotoxic effects on keratinocytes (HaCaT cell line). The results showed no decrease in cell viability after exposure to various concentrations of the extract, with cell viability increasing to approximately 100% at concentrations up to 100 μg/mL. This indicates that oleuropein-enriched extracts are non-cytotoxic to skin cells, even at relatively high concentrations [18].

Further evaluation of oleuropein derivatives from olive oil on human fibroblasts confirmed the safety of these phenolic compounds. The study demonstrated that the consumption and application of oleuropein derivatives did not compromise cell viability, highlighting their suitability for use in cosmetic formulations [19].

Overall, these findings underscore the potential of oleuropein as a safe and effective ingredient in cosmeceuticals. Its incorporation into skincare products offers multiple benefits, including antioxidant protection, enzyme inhibition, and photoprotection, without posing cytotoxic risks. This makes oleuropein an attractive option for developing innovative, sustainable cosmetic products that cater to the growing demand for natural and effective skincare solutions. By ensuring the safety and efficacy of oleuropein and its derivatives, the cosmetic industry can confidently harness these bioactive compounds to create products that deliver both health and beauty benefits to consumers.

4.4 Oleuropein and skin aging in cosmetic formulations

Oleuropein, a potent bioactive compound found in olive leaves and olive oil, plays a significant role in combating skin aging. It can inhibit and even reverse the signs of aging by reducing inflammation and oxidative damage, which in turn stimulates dermal reconstruction. Oleuropein, along with other polyphenols present in olive leaf extract, offers a range of beneficial properties for cosmetic formulations, including antioxidant, antibacterial, anti-hyaluronidase, anti-pigmentation, and wound healing effects. These properties make it an effective ingredient for anti-wrinkle treatments [20].

The anti-aging effects of olive leaf extract-containing cream were investigated by Wanitphakdeedecha et al. in a study involving 36 participants with photoaged skin. After applying the cream twice daily for 2 months, participants exhibited statistically significant improvements in skin texture, hydration, wrinkle reduction, and decreased melanin production. The study concluded that creams containing olive leaf extract possess notable anti-aging and rejuvenation properties, making them a valuable addition to cosmetic formulations [21].

A recent study further explored the use of byproducts such as olive leaf extracts in cosmetic creams. These extracts were obtained through supercritical fluid extraction, a method known for its efficiency and eco-friendliness. The study demonstrated that these extracts exhibited approximately 25% antioxidant activity and were non-cytotoxic to keratinocyte cells at concentrations up to 4% v/v within 24 hours. Incorporating these extracts into cosmetic formulations maintained product stability under various storage conditions, as confirmed by Turbiscan analyses. The safety of the final creams was validated through in vivo testing on human volunteers, which showed similar trans-epidermal water loss and erythematous index variations compared to the negative control. Additionally, rheological analyses indicated that the creams had suitable spreadability and pseudoplastic profiles, with only a slight decrease in viscosity at higher extract concentrations. This research underscores the potential of utilizing byproduct resources and supercritical fluid extraction to create safe, effective, and environmentally friendly cosmetic products. By harnessing the properties of oleuropein, the cosmetic industry can develop sustainable cosmeceutical formulations that offer real benefits to combat skin aging [17].

The scientific exploration of oleuropein's benefits also highlights its ability to modulate key biochemical pathways involved in skin health. Its antioxidant capacity helps neutralize free radicals, which are major contributors to the aging process. By reducing oxidative stress, oleuropein aids in maintaining skin integrity and elasticity. Furthermore, its anti-inflammatory properties help mitigate the chronic low-grade inflammation often associated with aging, thereby promoting a more youthful appearance.

As consumer demand for natural and effective skincare solutions grows, oleuropein-enriched formulations represent a promising frontier in the development of next-generation cosmetic products. This innovative approach not only supports the trend toward sustainability but also leverages the powerful bioactivities of oleuropein to deliver enhanced skin benefits.

4.5 The other biological benefits of oleuropein in skin health and cosmeceuticals

In recent years, there has been a growing interest in utilizing plant extracts in anti-aging cosmetic products, primarily for their antioxidant properties. These extracts help modulate the biochemical consequences of oxidative stress on the skin, which is a major contributor to aging. In the field of cosmeceuticals, antioxidants are powerful and innovative ingredients, and the application of exogenous antioxidants can effectively reduce the effects of free radicals. Oxidative stress occurs when the balance between reactive oxygen species (ROS) and antioxidants is disrupted, often due to the overproduction of ROS. This imbalance is particularly pronounced when the skin is exposed to ultraviolet (UV) radiation, where both UVB (280–320 nm) and UVA (320–400 nm) rays increase ROS production. During the photoaging process, UVB damage leads to skin wrinkling and pigmentation compared to unexposed skin [22].

The antioxidant activity of oleuropein, a bioactive compound derived from olive leaves and olive oil, is a prime example of its functional benefits in cosmeceuticals. Oleuropein exhibits a complex chemical structure that includes a phenolic hydroxyl group, a secoiridoid structure with an ester linkage, and an ortho-dihydroxyphenyl moiety (catechol structure). These structural features contribute to its antioxidant activity by donating hydrogen atoms to neutralize free radicals, thereby enhancing its ability to scavenge these radicals. Furthermore, the secoiridoid moiety, upon hydrolysis, forms simpler phenolic compounds that retain antioxidant properties. According to Japon-Lujan et al., olive leaves exhibit the highest antioxidant activity among the various parts of the olive tree, with compounds such as oleuropein, hydroxytyrosol, apigenin, caffeic acid, and other polyphenols supporting these findings [23].

Oleuropein has been shown to significantly mitigate skin damage and the incidence of skin tumors caused by long-term UVB irradiation in animal models, such as hairless mice. Studies have demonstrated that oleuropein reduces skin thickness, improves skin elasticity, and decreases skin carcinogenesis. Additionally, oleuropein inhibits the expression of matrix metalloproteases (MMPs), which are enzymes that degrade extracellular matrix proteins and activate proinflammatory cytokines like tumor necrosis factor (TNF)-alpha and interleukin (IL)-1 beta [24].

Our previous research demonstrated the antioxidant, antibacterial, and anti-aging effects of a lotion containing oleuropein, as well as its inhibitory impact on the enzymes tyrosinase, lipoxygenase, and hyaluronidase. In the same study, we also explored oleuropein's synergistic potential when formulated with other cosmeceuticals like *Helichrysum italicum* and kumquat essential oils. The lotion containing the combination of oleuropein and *H. italicum* exhibited the highest antioxidant and antimicrobial properties. Meanwhile, the oleuropein and kumquat combination showed the highest inhibitory effect on the enzymes: hyaluronidase, lipoxygenase, and tyrosinase. Oleuropein's water-soluble and hydrophilic properties enable it to penetrate the skin barrier easily, making it an ideal active ingredient in cosmetic formulations. Thus, oleuropein serves as a valuable source of active ingredients for cosmeceutical formulations [25].

According to Paulina et al., a 22.2% oleuropein-containing extract of olive leaves protects human fibroblast cells by reducing the overproduction of ROS in UVA-induced DNA damage. Oleuropein significantly inhibits both intrinsic and extrinsic apoptotic signaling pathways and increases the viability of fibroblasts by preventing apoptosis [26].

In a study by Katsiki et al., fibroblasts treated with oleuropein showed an extended lifespan, delayed aging, reduced intracellular ROS levels, and decreased amounts of oxidized proteins [27]. Li et al. demonstrated that oleuropein and its metabolite hydroxytyrosol exert synergistic cytoprotective effects by reducing ROS levels in human dermal fibroblasts. They also found that oleuropein and hydroxytyrosol exhibited moderate anti-elastase and anti-collagenase effects [28]. Elastase is a serine protease that breaks down elastin, a protein that provides elasticity to connective tissue, while collagenases are enzymes that degrade collagen, an essential component of the extracellular matrix.

In an in vitro study by Allaw et al., oleuropein was delivered to skin cells using collagen-enriched transferosomes, glycerosomes, and glytransferosomes, and its potential for wound healing on wounded fibroblast cell layers was tested. The results showed that when oleuropein (87%) was incorporated into these formulated vehicles, it remained highly stable for 4 months of storage. These vesicles increased cell viability against ROS and nitric oxide damage by 100%, resulting in enhanced fibroblast proliferation and wound area regeneration [20].

Another study explored the loading of oleuropein into chitosan-alginate microspheres, and the anti-inflammatory activity of the released proinflammatory cytokines, as well as the endogenous antioxidant content was measured. The results indicated that the microencapsulation of oleuropein significantly improved its anti-inflammatory and antioxidant capabilities in treated Lipopolysaccharide (LPS)/ human skin fibroblast cells [29]. Microencapsulation, a process where very fine solid or liquid particles are enclosed by a membrane, is increasingly applied in the cosmetic industry to enhance the stability and efficacy of active compounds.

5. Molecular mechanism of oleuropein action on skin health

5.1 Antioxidant and anti-inflammatory mechanisms

Oleuropein impacts skin health primarily through its potent antioxidant and anti-inflammatory mechanisms. As an antioxidant, oleuropein neutralizes reactive oxygen species (ROS) and other free radicals generated by ultraviolet (UV) radiation, pollution, and other environmental stressors. By scavenging these free radicals, oleuropein protects skin cells from oxidative damage, which can lead to premature aging, loss of collagen, and the development of fine lines and wrinkles.

Oleuropein also exhibits anti-inflammatory properties that further benefit skin health. It inhibits the expression of proinflammatory cytokines and enzymes, such as interleukin-1 (IL-1) and tumor necrosis factor-alpha (TNF-α), which are involved in the inflammatory response. By modulating these inflammatory pathways, oleuropein reduces redness, swelling, and irritation, which is particularly beneficial for conditions such as acne and rosacea. Moreover, oleuropein can enhance skin barrier function by stimulating the production of ceramides and other lipids, which helps retain moisture and protect against external irritants. These combined antioxidant and anti-inflammatory mechanisms make oleuropein an essential component in skin care, offering protection against environmental damage while soothing and enhancing the skin's natural defense systems. Maria et al. investigated the potential effects of oleuropein on IL-1β-induced production of inflammatory mediators and oxidative stress in the human synovial sarcoma cell line (SW982). They found that oleuropein exerted anti-inflammatory and antioxidant effects via down-regulation of NF-κB signaling pathways, reduced COX-2 activity, and decreased IL-6 and TNF-α cytokines [30].

5.2 Collagen preservation mechanisms

Oleuropein enhances collagen production through several mechanisms, including its antioxidant properties and its ability to modulate the activity of matrix metalloproteinases (MMPs). Oleuropein's antioxidant activity helps neutralize reactive oxygen species (ROS), which are known to damage skin cells and accelerate the degradation of collagen. By reducing oxidative stress, oleuropein helps maintain the structural integrity of skin cells and promotes a favorable environment for collagen synthesis. Additionally, oleuropein has been shown to stimulate the expression of collagen genes, enhancing the production of types I and III collagen, which are critical for maintaining skin strength and elasticity.

Furthermore, oleuropein impacts collagen metabolism by modulating the activity of matrix metalloproteinases (MMPs), particularly MMP-1 (collagenase-1), MMP-3 (stromelysin-1), and MMP-9 (gelatinase B), which are enzymes that break down collagen and other extracellular matrix components. Excessive MMP activity, often triggered by UV exposure and inflammatory processes, leads to increased collagen degradation and contributes to skin aging. Oleuropein has been shown to inhibit MMP activity by downregulating their expression and reducing the activation of signaling pathways such as mitogen-activated protein kinase (MAPK) and nuclear factor-kappa B (NF-κB), which are involved in MMP gene expression. By suppressing MMP activity, oleuropein not only reduces collagen breakdown but also promotes the maintenance and repair of the extracellular matrix, thereby supporting skin elasticity and firmness. This dual action of enhancing collagen synthesis and inhibiting collagen degradation underscores oleuropein's potential as a valuable ingredient in skin care and anti-aging formulations [28].

5.3 Anti-pigmentation mechanisms

Oleuropein promotes anti-pigmentation potential primarily by inhibiting the activity of tyrosinase, a key enzyme involved in melanogenesis, the process responsible for melanin production in the skin. Tyrosinase catalyzes the initial steps in the conversion of the amino acid tyrosine into melanin, specifically the hydroxylation of tyrosine to L-DOPA and the subsequent oxidation of L-DOPA to dopaquinone, which ultimately leads to the synthesis of eumelanin and pheomelanin. Oleuropein exerts its anti-pigmentation effects by directly binding to the active site of tyrosinase, thereby inhibiting its catalytic activity. This inhibition reduces the production of melanin, resulting in lighter skin pigmentation and preventing hyperpigmentation disorders such as melasma, age spots, and post-inflammatory hyperpigmentation.

In addition to direct inhibition of tyrosinase, oleuropein may also impact pigmentation through its antioxidant properties. By scavenging reactive oxygen species (ROS) and reducing oxidative stress, oleuropein can prevent the activation of signaling pathways that upregulate tyrosinase expression, such as the microphthalmia-associated transcription factor (MITF) pathway. MITF is a critical regulator of melanocyte function and melanogenesis, and its expression can be induced by oxidative stress and UV exposure. By reducing ROS levels, oleuropein may downregulate MITF expression, further decreasing tyrosinase activity and melanin production. Thus, oleuropein's ability to inhibit tyrosinase activity and modulate oxidative stress-related signaling pathways contributes to its potential as an anti-pigmentation agent in skin care formulations [31].

6. Comparing oleuropein and vitamin C efficacy in cosmeceutical applications

Vitamin C has long been a cornerstone in cosmeceuticals, valued for its potent antioxidant properties and ability to promote skin health. Used since the early twentieth century, it has been integral in formulations aimed at brightening the skin, reducing signs of aging, and enhancing overall radiance. Its established efficacy in neutralizing free radicals and supporting collagen synthesis has cemented its role as a key active ingredient in skincare products [32].

6.1 Antioxidant properties

Vitamin C is a potent antioxidant, effectively neutralizing reactive oxygen species (ROS) and regenerating other antioxidants like vitamin E. Its strength in protecting skin cells from oxidative damage is well-established. Oleuropein also offers significant antioxidant activity, comparable to vitamin C. However, its additional benefits include reducing oxidative stress through multiple pathways, which can enhance overall skin protection. Oleuropein's antioxidant properties can be considered stronger in the context of its broader impact on cellular oxidative stress.

6.2 Collagen production

Vitamin C is crucial for collagen synthesis, acting as a cofactor for prolyl and lysyl hydroxylases, which are essential for collagen maturation. Its role in enhancing collagen production is well-documented and substantial.

Oleuropein supports collagen production by inhibiting matrix metalloproteinases (MMPs) that degrade collagen. While vitamin C directly stimulates collagen synthesis, oleuropein's strength lies in its ability to preserve existing collagen and prevent its breakdown, offering a unique advantage in maintaining skin resilience and firmness.

6.3 Anti-inflammatory effects

Vitamin C reduces inflammation by modulating cytokine levels and inhibiting inflammatory pathways, providing effective relief from redness and swelling. Oleuropein also has anti-inflammatory properties and may offer a more comprehensive effect by reducing oxidative stress and modulating inflammatory pathways. Oleuropein can be considered stronger in terms of its broader impact on inflammation, providing enhanced soothing effects and reducing skin irritation.

6.4 Anti-pigmentation

Vitamin C inhibits tyrosinase, the enzyme responsible for melanin production, making it effective in reducing hyperpigmentation and evening out skin tone. Oleuropein also inhibits tyrosinase but may offer additional benefits due to its antioxidant properties, which help reduce oxidative stress that can exacerbate pigmentation issues. Oleuropein's strength in anti-pigmentation lies in its combined ability to inhibit melanin production and reduce oxidative stress, potentially offering more comprehensive skin tone correction.

6.5 Overall skin benefits

Vitamin C is well-regarded for its ability to brighten skin, reduce fine lines and wrinkles, and enhance radiance, making it a cornerstone of many skincare routines. Oleuropein provides complementary benefits, including superior collagen preservation, enhanced anti-inflammatory effects, and additional antioxidant protection. Oleuropein's unique advantage lies in its multifaceted approach to skin health, addressing multiple aspects of skin aging and damage simultaneously.

As a result, while both oleuropein and vitamin C offer significant benefits, oleuropein's unique advantages include its broader antioxidant impact, stronger collagen preservation, enhanced anti-inflammatory effects, and comprehensive anti-pigmentation benefits. This makes oleuropein a valuable addition to cosmeceutical formulations, offering complementary benefits to those provided by vitamin C.

7. Optimal dosage of oleuropein in cosmetic formulations and influencing factors

7.1 Optimal dosage

The optimal dosage of oleuropein in cosmetic formulations generally ranges from 0.5–2%. At these concentrations, oleuropein effectively delivers its antioxidant, anti-inflammatory, and collagen-preserving benefits without causing irritation. This range ensures that oleuropein can exert its positive effects on skin health while maintaining safety and compatibility with various skin types [33].

7.2 Factors influencing efficacy

7.2.1 Formulation type

Delivery Systems: The effectiveness of oleuropein can be influenced by the delivery system used. Advanced delivery systems like liposomes, nanoparticles, or microemulsions can enhance its penetration and stability, improving its bioavailability and efficacy in the skin.

Stability: Oleuropein's stability in different formulations (e.g., creams and serums) can impact its efficacy. Formulations that protect oleuropein from oxidation and degradation are crucial for maintaining its effectiveness over time.

7.2.2 pH levels

pH Compatibility: The pH of the cosmetic formulation can affect the stability and efficacy of oleuropein. Optimal pH ranges that align with the skin's natural pH (around 4.5 to 5.5) are important to ensure that oleuropein remains stable and active.

7.2.3 Interaction with other ingredients

Synergistic Effects: Oleuropein may interact with other active ingredients, such as vitamins C or E, enhancing overall efficacy through synergistic effects. Conversely, certain ingredients might affect oleuropein's stability or bioavailability.

Incompatibility: Potential chemical interactions with other components in the formulation can influence oleuropein's performance, necessitating careful formulation design to avoid negative interactions.

7.2.4 Application method

Frequency and Duration: The effectiveness of oleuropein can also depend on the frequency of application and duration of use. Regular and consistent application is necessary to achieve and maintain visible results.

7.2.5 Skin type

Individual Variability: Efficacy can vary based on individual skin types and conditions. For instance, oleuropein may be more effective for certain skin concerns (e.g., oxidative stress and inflammation) but may require adjustments in concentration or formulation for optimal results across different skin types.

In summary, the optimal dosage of oleuropein in cosmetic formulations typically ranges from 0.5 to 2%. Its efficacy can be influenced by factors such as formulation type, pH levels, interactions with other ingredients, application methods, and individual skin characteristics. Careful consideration of these factors is essential to maximize the benefits of oleuropein in skincare products.

8. Innovative delivery systems for enhanced oleuropein penetration and bioavailability

8.1 Nanoparticles and nanocarriers

Liposomes: These lipid-based vesicles encapsulate oleuropein, enhancing its stability and facilitating its delivery to deeper skin layers. Liposomes can improve oleuropein's penetration through the stratum corneum, increasing its bioavailability.

Nanospheres and Nanocapsules: Using materials such as poly (lactic-co-glycolic acid) (PLGA) or polylactic acid (PLA), these carriers provide a controlled release and sustained delivery of oleuropein, improving its efficacy over time.

8.2 Microemulsions

Oil-in-Water (O/W) and Water-in-Oil (W/O) Microemulsions: These stable, transparent systems enhance oleuropein's solubility and permeability through the skin barrier. Microemulsions facilitate deeper skin penetration and ensure uniform distribution of oleuropein.

8.3 Hydrogels

Smart Hydrogels: These are responsive materials that release oleuropein in response to environmental triggers such as pH changes or temperature. Hydrogels can offer targeted delivery and prolonged release of oleuropein, enhancing its effectiveness.

8.4 Microneedle patches

Dissolving Microneedles: Microneedle patches create micro-channels in the skin, allowing for direct delivery of oleuropein into the dermis. The dissolving nature of these microneedles ensures that oleuropein is effectively absorbed without causing discomfort.

8.5 Encapsulation in cyclodextrins

Cyclodextrin Complexes: Encapsulating oleuropein in cyclodextrins can enhance its stability and solubility. This method improves oleuropein's ability to permeate the skin and ensures its effective release at the target site.

8.6 Electroporation

Pulsed Electric Fields: This technique temporarily disrupts the skin barrier using electric fields, allowing oleuropein to penetrate more deeply into the skin. It offers a method for improving the bioavailability of oleuropein without invasive procedures.

These innovative delivery systems can significantly enhance the penetration, stability, and overall efficacy of oleuropein in cosmeceutical applications, ensuring that its beneficial effects are maximized for skin health.

9. Future research and study limitations

To substantiate the efficacy of oleuropein in enhancing skin health within cosmeceuticals, there is a critical need for more rigorous clinical trials. Current studies, while promising, often involve small sample sizes or lack robust control measures, limiting the generalizability and reliability of their findings. Future research should focus on large-scale, well-designed clinical trials that include diverse populations to better understand oleuropein's effects on various skin conditions and its long-term benefits.

Existing studies on oleuropein's role in skin health face several limitations. Many studies rely on in vitro or animal models, which may not fully replicate human skin responses. Additionally, variations in formulation concentrations and application methods can impact outcomes, making it challenging to establish standardized recommendations. There is also a need for research into the optimal dosage, delivery mechanisms, and potential interactions with other active ingredients. Addressing these gaps will help validate oleuropein's efficacy and safety, providing a clearer foundation for its use in cosmeceuticals.

10. Conclusion

In conclusion, the research into oleuropein's role in skin health emphasizes its potential as a valuable compound for cosmeceutical applications. As outlined in this chapter, the skin serves as a physical barrier and protective organ, requiring both nutritional and topical support to maintain its integrity and liveliness. Cosmeceuticals, placed at the intersection of cosmetics and pharmaceuticals, offer a promising avenue for delivering bioactive compounds like oleuropein across skin layers.

Oleuropein, a phenolic compound derived from olive leaves and oil, exhibits a numerous beneficial property relevant to skin health including antioxidant, anti-pigmentation, anti-collagenase, anti-elastase, anti-hyaluronidase, anti-inflammation, and wound healing effects. These properties of oleuropein enhance collagen, elastin, and hyaluronic acid production in dermis, which provide skin strength and hydration and express its role in supporting skin structure and function.

The integration of oleuropein into cosmeceuticals formulations presents an innovative approach to skincare, offering consumers products that not only improve the esthetic appearance of the skin but also contribute to its overall health. Future *in vitro*, *in vivo*, and clinical research in this area could lead to more effective formulations and delivery systems, representing the benefits of oleuropein and its metabolites like hydroxytyrosol.

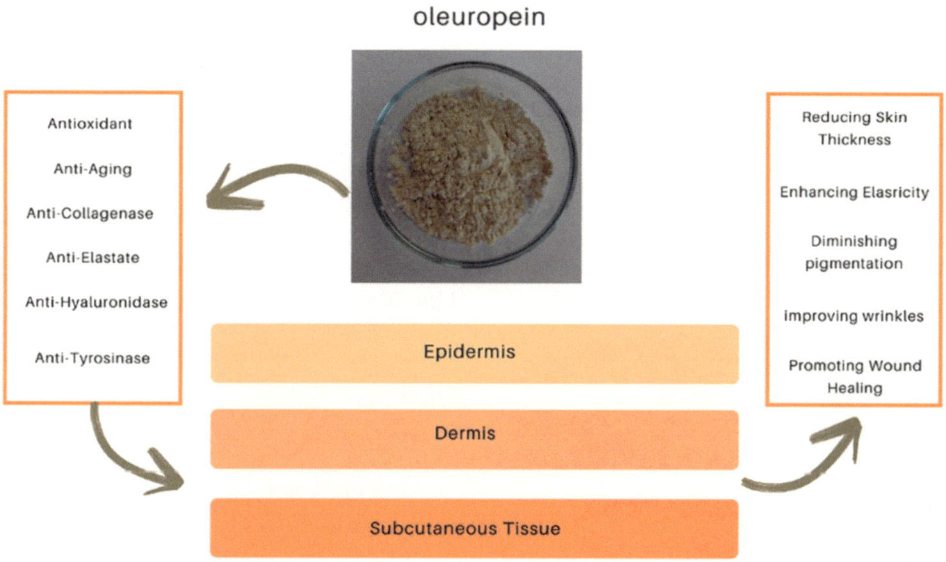

Figure 1.
The role of oleuropein on cosmeceuticals.

As the demand for natural products continues to grow, oleuropein's role in cosmeceuticals is likely to expand, offering new possibilities for consumers seeking healthier, more luminous skin (**Figure 1**).

Acknowledgements

We would also like to acknowledge Uskudar University for providing the necessary facilities and study environment that enabled us to conduct this study.

Author details

Shirin Tarbiat
Uskudar University, Istanbul, Turkey

*Address all correspondence to: shirin.tarbiat@uskudar.edu.tr

References

[1] Michalak M, Pierzak M, Kręcisz B, Suliga E. Bioactive compounds for skin health: A review. Nutrients. 2021;**13**(1):203. DOI: 10.3390/nu13010203

[2] Trojahn C, Dobos G, Blume-Peytavi U, Kottner J. The skin barrier function: Differences between intrinsic and extrinsic aging. Giornale Italiano di Dermatologia e Venereologia. 2015;**150**(6):687-692

[3] Park SH, Park J, Lee M, Jun W, Kim J, Geum J, et al. Wheat ceramide powder mitigates ultraviolet B-induced oxidative stress and photoaging by inhibiting collagen proteolysis and promoting collagen synthesis in hairless mice. Preventive Nutrition and Food Science. 2023;**28**(4):418-426. DOI: 10.3746/pnf.2023.28.4.418

[4] Reszko AE, Berson D, Lupo MP. Cosmeceuticals: Practical applications. Dermatologic Clinics. 2009;**27**(4):401-416. DOI: 10.1016/j.det.2009.08.005

[5] Hernández-Fernández A, Garrido Y, Iniesta-López E, Pérez de los Ríos A, Quesada-Medina J, Hernández-Fernández FJ. Recovering polyphenols in aqueous solutions from olive mill wastewater and olive leaf for biological applications. PRO. 2023;**11**(9):2668. DOI: 10.3390/pr1109

[6] Bucciantini M, Leri M, Nardiello P, Casamenti F, Stefani M. Olive polyphenols: Antioxidant and anti-inflammatory properties. Antioxidants. 2021;**10**(7):1044. DOI: 10.3390/antiox1007

[7] Mohd-Setapar SH, John CP, Mohd-Nasir H, Azim MM, Ahmad A, Alshammari MB. Application of nanotechnology incorporated with natural ingredients in natural cosmetics. Cosmetics. 2022;**9**(6):110. DOI: 10.3390/cosmetics9060110

[8] Kolarsick PAJ, Kolarsick MA, Goodwin C. Anatomy and physiology of the skin. Journal of the Dermatology Nurses' Association. 2011;**3**(4):203-213. DOI: 10.1097/JDN.0b013e3182274a98

[9] Wong QYA, Chew FT. Defining skin aging and its risk factors: A systematic review and meta-analysis. Scientific Reports. 2021;**11**(1):22075. DOI: 10.1038/s41598-021-01573-z

[10] Lau M, Mineroff Gollogly J, Wang JY, Jagdeo J. Cosmeceuticals for antiaging: A systematic review of safety and efficacy. Archives of Dermatological Research. 2024;**316**(5):173. DOI: 10.1007/s00403-024-02908-2

[11] Hoang HT, Moon J-Y, Lee Y-C. Natural antioxidants from plant extracts in skincare cosmetics: Recent applications, challenges and perspectives. Cosmetics. 2021;**8**(4):106. DOI: 10.3390/cosmetics8040106

[12] Gomes C, Silva AC, Marques AC, Sousa Lobo J, Amaral MH. Biotechnology applied to cosmetics and aesthetic medicines. Cosmetics. 2020;**7**(2):33. DOI: 10.3390/cosmetics7020033

[13] Khalil AA, Rahman MM, Rauf A, Islam MR, Manna SJ, Khan AA, et al. Oleuropein: Chemistry, extraction techniques and nutraceutical perspectives-An update. Critical Reviews in Food Science and Nutrition. 2023. pp. 1-22. DOI: 10.1080/10408398.2023.2218495

[14] Dauber C, Parente E, Zucca MP, Gámbaro A, Vieitez I. Olea europea

and by-products: Extraction methods and cosmetic applications. Cosmetics. 2023;**10**(4):112. DOI: 10.3390/cosmetics10040112

[15] Otero P, Garcia-Oliveira P, Carpena M, Barral-Martinez M, Chamorro F, Echave J, et al. Applications of by-products from the olive oil processing: Revalorization strategies based on target molecules and green extraction technologies. Trends in Food Science & Technology. 2021;**116**:1084-1104. DOI: 10.1016/j.tifs.2021.09.007

[16] Rishmawi S, Haddad F, Dokmak G, Karaman R. A comprehensive review on the anti-cancer effects of oleuropein. Life. 2022;**12**(8):1140. DOI: 10.3390/life12081140

[17] Nunes A, Gonçalves L, Marto J, Martins AM, Silva AN, Pinto P, et al. Investigations of olive oil industry by-products extracts with potential skin benefits in topical formulations. Pharmaceutics. 2021;**13**(4):465. DOI: 10.3390/pharmaceutics13040465

[18] MdlL C-G, Pinto D, Delerue-Matos C, Rodrigues F. Olive fruit and leaf wastes as bioactive ingredients for cosmetics—A preliminary study. Antioxidants. 2021;**10**(2):245. DOI: 10.3390/antiox10020245

[19] González-AcedoA,Ramos-TorrecillasJ, Illescas-Montes R, Costela-Ruiz VJ, Ruiz C, Melguizo-Rodríguez L, et al. The benefits of olive oil for skin health: Study on the effect of hydroxytyrosol, tyrosol, and oleocanthal on human fibroblasts. Nutrients. 2023;**15**(9):2077. DOI: 10.3390/nu15092077

[20] Allaw M, Manca ML, Gómez-Fernández JC, Pedraz JL, Terencio MC, Sales OD, et al. Oleuropein multicompartment nanovesicles enriched with collagen as A natural strategy for the treatment of skin wounds connected with oxidative stress. Nanomedicine. 2021;**16**(26):2363-2376. DOI: 10.2217/nnm-2021-0197

[21] Wanitphakdeedecha R, Ng JNC, Junsuwan N, Phaitoonwattanakij S, Phothong W, Eimpunth S, et al. Efficacy of olive leaf extract–containing cream for facial rejuvenation: A pilot study. Journal of Cosmetic Dermatology. 2020;**19**(7):1662-1666. DOI: 10.1111/jocd.13457

[22] Wei M, He X, Liu N, Deng H. Role of reactive oxygen species in ultraviolet-induced photodamage of the skin. Cell Division. 2024;**19**(1):1. DOI: 10.1186/s13008-024-00107-z

[23] Japón-Luján R, Luque-Rodríguez JM, Luque de Castro MD. Dynamic ultrasound-assisted extraction of oleuropein and related biophenols from olive leaves. Journal of Chromatography A. 2006;**1108**(1):76-82. DOI: 10.1016/j.chroma.2005.12.106

[24] Kimura Y, Sumiyoshi M. Olive leaf extract and its main component oleuropein prevent chronic ultraviolet B radiation-induced skin damage and carcinogenesis in hairless mice. The Journal of Nutrition. 2009;**139**(11):2079-2086. DOI: 10.3945/jn.109.104992

[25] Tarbiat S, Yener FG, Kashefifahmian A, Mohseni AR. Antiaging effects of oleuropein combined with *Helichrysum italicum* or kumquat essential oils in cosmetic lotions. Current Topics in Nutraceutical Research. 2022;**20**(2):352-359. DOI: 10.37290/ctnr2641-452X.20:352-359

[26] Machała P, Liudvytska O, Kicel A, Dziedzic A, Olszewska MA, Żbikowska HM. Valorization of the photo-protectivepotentialofthephytochemically standardized olive (*Olea europaea* L.) leaf

extract in UVA-irradiated human skin fibroblasts. Molecules. 2022;**27**(16):5144. DOI: 10.3390/molecules27165144

[27] Katsiki M, Chondrogianni N, Chinou I, Rivett AJ, Gonos ES. The olive constituent oleuropein exhibits proteasome stimulatory properties in vitro and confers life span extension of human embryonic fibroblasts. Rejuvenation Research. 2007;**10**(2):157-172. DOI: 10.1089/rej.2006.0513

[28] Li H, He H, Liu C, Akanji T, Gutkowski J, Li R, et al. Dietary polyphenol oleuropein and its metabolite hydroxytyrosol are moderate skin permeable elastase and collagenase inhibitors with synergistic cellular antioxidant effects in human skin fibroblasts. International Journal of Food Sciences and Nutrition. 2022;**73**(4):460-470. DOI: 10.1080/09637486.2021.1996542

[29] Hendawy OM, Al-Sanea MM, Mohammed Elbargisy R, Ur Rahman H, Hassan YA, Elshaarawy RFM, et al. Alginate-chitosan-microencapsulated tyrosols/oleuropein-rich olive mill waste extract for lipopolysaccharide-induced skin fibroblast inflammation treatment. International Journal of Pharmaceutics. 2023;**643**:123260. DOI: 10.1016/j.ijpharm.2023.123260

[30] Castejón ML, Rosillo MÁ, Montoya T, González-Benjumea A, Fernández-Bolaños JM, Alarcón-de-la-Lastra C. Oleuropein down-regulated IL-1β-induced inflammation and oxidative stress in human synovial fibroblast cell line SW982. Food & Function. 2017;**8**(5):1890-1898

[31] Wang H, Chen J, Hu J, Si J, Xie Y, Wei J, et al. Tyrosinase inhibitor screened from Olea europaea L. leaves: Identification, molecular docking analysis and molecular mechanisms. Industrial Crops and Products. 2024;**210**:118112. DOI: 10.1016/j.indcrop.2024.118112

[32] Kumar AV, Garg VK, Buttar HS, Choudhary S, Sharma S, Grover A. The protective role of vitamins in skincare and cosmeceutical products: Mechanisms involved to re-engineer the skin towards a healthy state. In: Shah AK, Tappia PS, Dhalla NS, editors. Hydrophilic Vitamins in Health and Disease. Cham: Springer International Publishing; 2024. pp. 105-124

[33] El-Gogary RI, Ragai MH, Moftah N, Nasr M. Oleuropein as a novel topical antipsoriatic nutraceutical: Formulation in microemulsion nanocarrier and exploratory clinical appraisal. Expert Opinion on Drug Delivery. 2021;**18**(10):1523-1532

Chapter 4

Olive in Egypt: Cultural Practices for Sustainable Production

Ahmed Mohamed Taha and Hamdy El-Houssainy Khalifa

Abstract

Worldwide, olive is well known as an important economic and social crop. Historically, Egypt is one of the major producers of olive under arid and semi-arid conditions in the Mediterranean region under irrigated agriculture. The sustainability of olive production in Egypt faces several problems including water scarcity for agricultural sector, soil deterioration, increasing production cost, and climate variability. Therefore, it is important to evaluate the factors affecting olive production on a sustainable base and initiate mitigation/adaptation strategies to avoid and cope with the prevailing problems. The current chapter reviews the most recent studies on olive production practices, the main impacts of climate change on olive tree cultivation, and the possible mitigation/adaptation strategies against the potentially negative impacts of climate variability under Egyptian conditions.

Keywords: olive varieties and production, water amounts, crop coefficient, water productivity, good agricultural practices, climate variability, and adaptation strategies

1. Introduction

The agricultural sector, including forestry and fishing, is a key sector that significantly contributes to Egypt's Gross Domestic Product (GDP) by 11.83% and provides 28% of total employment [1, 2]. According to IMF [3], the Egyptian economy is the third highest GDP among African countries in the year 2021 and is the largest in North Africa.

The sustainability and environmental aspects of agricultural sector in Egypt are:

1.1 Land resources

Egypt has a total land area of approximately 1 million km^2. The agricultural land is nearly 4.03 million hectares in 2021, which corresponds to about 4.1% of the country's total land area. The agricultural land base consists of four main areas: (1) in the Nile Valley and Delta (Old Lands), (2) reclaimed from the desert since 1952 (New Lands), (3) rainfed areas in the north coast, and (4) several oases where groundwater is used for irrigation [4]. Geographically, the Old Lands represent the largest irrigated area in Egypt and are found in the Nile Valley and Delta, cover about 2.73 million ha. These lands, characterized by alluvial soils, are irrigated by traditional surface irrigation systems, which have a very low field water application efficiency of around 50%. The soil is mainly of clay texture with low organic matter content and available nitrogen due to the intensive agricultural practices. In addition, these soils will further suffer from salinization due to declining of water supplies with direct effect on irrigation water quantities and qualities as well as rising sea level [5]. Two problems occur on most of these lands: encroachment by non-agricultural uses and continued degradation of soil fertility [6]. The New Lands cover about 1.05 million ha and are located on the Old Lands' fringes and in the deserts. The soils in the New Lands are deprived of macro- and micro-nutrients, very low in organic matter contents, high pH, and have poor physical and hydro-physical characteristics. In many areas, the soils include high $CaCO_3$ and salinity contents as well as gypsum is present in some areas. Hard pans that are formed at different depths in the soil profile due to the presence of many cementing agents represent the main physical constraints in the New Lands. Also, the risk of erosion is high [5, 7]. All newly developed agricultural areas of the New Lands use pressurized irrigation systems [8]. The rainfed areas represent 0.17 million ha of land located in the north coastal areas, where rainfall varies annually between 100 and 200 mm. The oases cover a total area of 40,000 ha, which is irrigated mainly from the groundwater. The soils in the oasis are mainly alluvial, sand, and calcareous [9].

1.2 Farming systems

In Egypt, small farms (<0.38 ha) represent the major part of agricultural sector in which the farmers apply outdated practices that do not meet the documented standards. Therefore, the growers face high production costs, low yields, increasing salinity of soil and water resources, poor soil fertility, and restricted marketing opportunities [10].

1.3 Water resources

Egypt is characterized by a dry climate, scarce rainfall, and its water supplies are uncertain. Water resources in Egypt are limited to either conventional (Nile River inflow, groundwater, and rainfall) or non-conventional (seawater desalination and the use of wastewater). Consequently, the main task is to solve the problems raised by the increasing gap between the restricted water resources and the increasing requirements for domestic and irrigation water [11–13].

1.3.1 Conventional water

1.3.1.1 River Nile

Egypt relies on the Nile for its water needs. The length of River Nile is 6695 km and covers about 2.9 million km^2 in ten African countries. About 55.5 (BCM/yr.) are received from the 1660 BCM/yr. of water that falls inside the Nile Basin. Almost 86% of the Nile's flows from the Ethiopian Plateau (Blue Nile 59%, Sobat 14%, and Atbara 13%), while 14% is the contribution from the Equatorial Lakes (White Nile) [14].

1.3.1.2 Rainfall

Egypt is considered to be arid and hyper-arid country that receives very little rainfall. Rainwater is concentrated along the north coast of Egypt and decreases gradually to the south. The average yearly rainfall rate varies between 50 and 250 (mm). Rainfall increases toward the northeast, averaging 150 mm in Arish while reaching about 250 mm in Rafah. The total amount of rainwater is about one billion m^3 per year [14].

1.3.1.3 Groundwater

The groundwater is the second source of freshwater in Egypt and represents about 12% of the water supply. The renewable groundwater is primarily pumped from the Nile Valley, the Delta region, and desert reservoirs. This water is actually a component of the Nile's water resources. There are several challenges facing groundwater use including: (1) severe groundwater withdrawal from the aquifer systems, (2) changes in the groundwater properties (physical and chemical) compared to surface water owing to the negative impact of the prevailing environment, and (3) effect of surface activities, seepage of rainwater, and the use of irrigation and drainage water and other effluents. In northern Egypt, the aquifers are at risk due to seawater intrusion and saltwater seepage. Also, the groundwater is affected by the influence of climate change on recharge rates that decrease in proportion to a decrease in precipitation [15].

1.3.2 Non-conventional water

1.3.2.1 Reuse of agricultural water

In Egypt, water demand is larger than the conventional supply therefore the utilization of non-conventional water resources became essential. In Upper Egypt, drainage water returns to the Nile and is indirectly used for irrigation. In the Delta, reused water comes from overirrigation in the cultivated areas. This water includes leakage losses from the irrigation and drainage network. It is considered to be low-quality water because of its high salinity (ranging from 700 to more than 3000 ppm) and its mixing with drain water, which is often polluted [14].

1.3.2.2 Reuse of treated wastewater

This water is used for irrigation purposes considering that health requirements are met. The annual amount of treated water increased from 0.26 BCM/yr. in the early 1990s to about 0.6 BCM/yr. in 2000, eventually reaching about 3.7 BCM in 2017. It is used to irrigate non-food crops for humans and animals and to cultivate trees in the desert to produce wood [14].

1.3.2.3 Seawater desalination

Currently, Egypt has 90 seawater desalination plants in operation, with a total capacity of 1.3 BCM, costing 0.4 billion dollars. The daily capacity is 850×103 m^3. The Egyptian government aims to provide 6.4×106 m^3 per day of freshwater using the desalination technique [14].

1.4 Climate change

Significant challenges in agriculture farming are expected as a result of climate change. The unpredictable weather patterns, rising temperatures, and increased frequency of extreme events have the potential to disturb agricultural practices and threaten food security. In Egypt, climate change worsens the already constrained water and land resources in the coastal region by increasing the groundwater salinity and loss of cultivated agricultural area because of seawater intrusion, increasing water requirement of the crops due to increasing temperatures, and the more frequent extreme weather events [16, 17]. Results of a study based on The International Model for Policy Analysis of Agricultural Commodities and Trade (IMPACT) showed that climate change will reduce aggregate food production from −3% (2030) to −3.8% (2050) [18]. The decrease in food production will lead to an increase in prices and a decrease in per capita food consumption (kilocalories/capita/day) by about −1.7 and −3.8% during 2030 and 2050, respectively. The investigation concluded that, during the period from 2030 to 2050, there will be an increase, on average by 0.017%, in hunger, and millions of people will be at risk. The study recommended that adaptive measures be developed by Egyptian policymakers to consider the potential impacts of climate change that persuade losses in agricultural productivity [19, 20]. To cope with the negative impact of climate change, it is essential to breed crop varieties that are resistant to heat and drought, improve agricultural practices, and educate farmers to shift from crops that are highly affected by climate change to those that are tolerant. In this direction, major efforts are made to save 30% of irrigation water by implementing breeding programs to produce short-duration crop varieties (e.g., rice, wheat, corn, and barley). Also, the agricultural sector needs more investment to improve productivity [10, 21].

2. Olives in Egypt

2.1 Historical background

Olives are one of the oldest fruits in the world. Olive trees are the most suitable and best-adapted species to the Mediterranean-type climate. Climate changes and variability represent new challenges that are predicted to arise and threaten this traditional crop [22].

In Egypt, there is evidence documenting since the Pharaonic period. The olive tree in Egypt goes back to Dynasty XVIII (1570–1345 BC), and the documents revealed that Pharoah Ramses III (1197–1165 BC) promoted the cultivation of olive trees. The modern Egyptian olive cultivars are significantly distinct from the rest of the reference varieties. The cultivation of these varieties dates back to the first Persian rule over Egypt [23].

2.2 Olive production

Egypt is one of the top 10 olive producers in the world, with 5.5% of total world production in 2019. Table olive represents about 95% of Egypt's production. It is used for both domestic consumption and export. The International Olive Council (IOC) reported that, in 2019, Egypt was the world's largest producer of table olives with 25% of total world production. Most of olive production is located in the North-Western coastal region, Alexandria, North Sinai, and the Oases [24]. The harvested olive areas are distributed in the Old and New Lands as well as outside the Nile Valley and Delta (**Table 1** and **Figure 1**). The areas cultivated in the New Lands represented 94.4%, while the areas in the Old Lands represented 5.6% of the total harvested area.

The area cultivated by olives increased from 45,513 ha in 2000 to 112,657 ha in 2022, with a 147.5% increase in area (**Figure 2a**). Olive production has increased from 281,745 tons in 2000 to 1,137,075 tons in 2022, an increase of 303 percent (**Figure 2b**), and an increase in olive oil production from 8000 to 40,000 tons from 2000 to 2022, an increase of 400% (**Figure 2c**).

2.3 Olive varieties

There are more than thousand olive varieties from all over the world, each with its own unique quality, color, taste and oil characteristics. The most popular and widely cultivated local and international olive varieties used in Egypt are:

Location	Lower Egypt	Middle Egypt	Upper Egypt	Outside	Total
Old Land (ha)	1880.25	3335.71	1093.28	—	6309.24
New Land (ha)	35210.08	10954.2	1292.44	58891.18	106347.9
Total	37090.33	14289.91	2385.72	58891.18	112657.14

Table 1.
Harvested olive areas in Egypt during 2022.

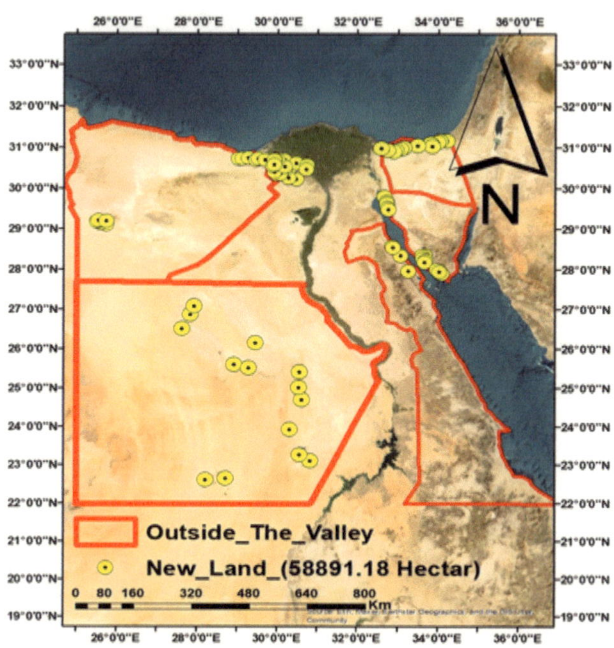

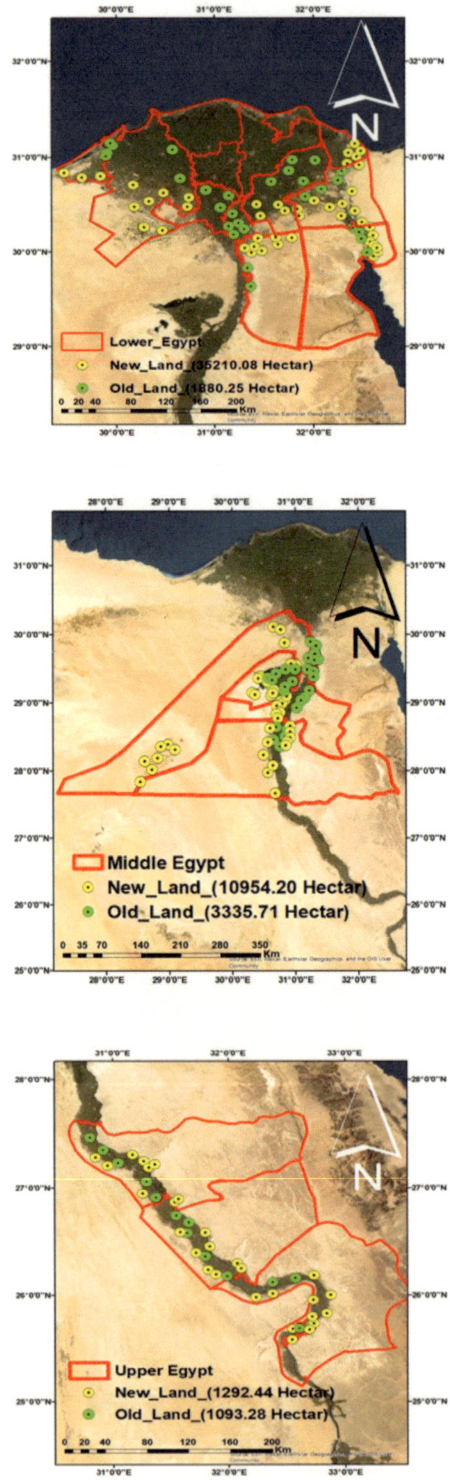

Figure 1.
Harvested olive areas in lower, middle, and upper Egypt and outside the Valley and Delta during 2022.

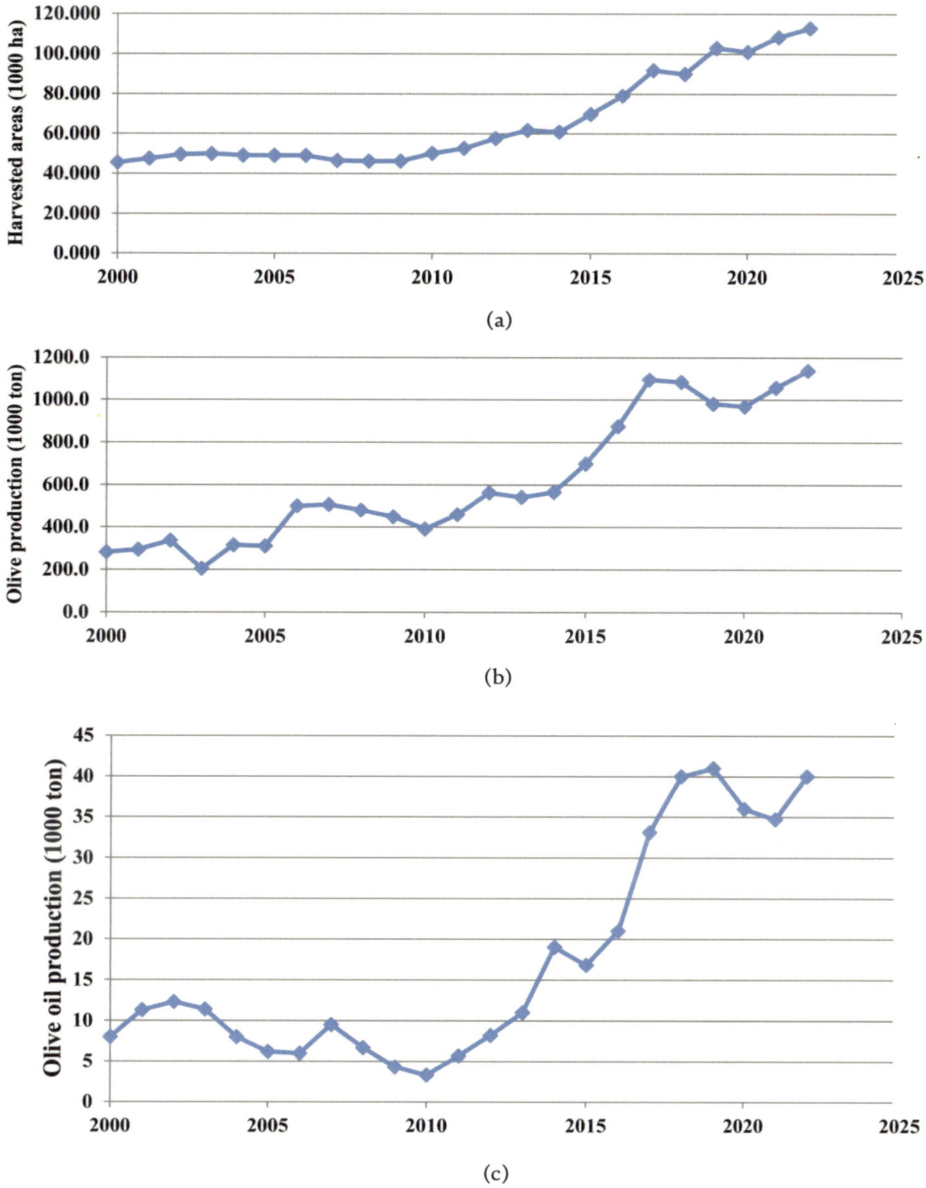

Figure 2.
Evolution of harvested area, total olive and oil production from 2000 to 2022.

2.3.1 Local varieties

Aggizi Shami: It is adapted to severe environmental conditions including high temperature, solar radiation, and deficit water. The variety has a high rooting ability and an intermediate bearing start with medium productivity. It has a large fruit size and freestone as well as a high flesh/stone ratio. The flesh is tasty, firm, and resistant to handling. The fruit has an almond-shape with medium weight (7–10 g). It is moderately susceptible to the olive fly. It is suitable for the production of green or stuffed olives due to its low oil content (7–9%). It is one of the most important table olive varieties in Egypt and accounts for 20% of olive area.

Aggizi Aksi: It has green medium fruits. Seed represents 10–15% of the fruit. It has an elongated shape with smaller size of 6–8 g. The rooting ability and productivity are high. The fruits are very heavy, with a medium flesh/stone ratio. It is excellent in processing. Initially, it was found on a limited scale, but it is now widely distributed in the country.

Hamed: It is very ancient in origin and comes from the Siwa Oasis. The variety is very hardy. The trees, which are large in size, may suffer from excessive fruit load. The rooting ability is good and has an intermediate bearing start. The fruit is large and very sensitive to damage during handling and transportation. It has a high fruits' flesh/stone ratio and yield is high and constant. The variety is tolerant to drought and salinity. It is cultivated in 6% of the olive area.

Maraki: The variety comes from the district of the Siwa Oasis. It is used for oil production. The fruit is very heavy with a medium flesh/stone ratio. Average fruit weight is 5.63–6.86 g, average fruit length/width ratio is 2.04–2.06, fruits had spherical to ovoid shape, and average oil percentage varies from 52.82 to 53.45. Its oil is very high in oleic acid content with medium bitterness. The crop intensity is very high with low rooting ability. The best harvest time is from November to December. Its distribution is currently limited, about 2% of current olive area and is expected to become Egypt's main oil cultivar.

Sewia: Average fruit weight 6.63–7.16 g, average length/width ratio for fruits is 1.35–1.42, fruits had ovoid shape, average stone weight of ripe fruits is 1.07–1.2 g, and average oil percentage varies from 46.22 to 50.86.

Toffahi: It is one of the local varieties which are widespread planting in Egypt. The fruit is large round, weighing 8–16 g, jagged nucleus stuck a little in the meat and make up 13% of the weight of the fruit. Percentage of oil is 5–7%. Fruits are used only in green pickling and it is matures early in late August until the end of September. It is sensitive to an excavator leg injury and worm green olive leaves. Fruits do not bear long-term conservation.

Wateken: It is an old variety initially from Siwa Oasis. Fruit weight is very high, with a medium flesh/stone ratio. It is mainly used for oil production. The variety has a medium rooting ability and crop intensity. The best harvest time is between October and December. It is locally distributed in Siwa Oasis area and accounts for 5% of the olive area.

2.3.2 International varieties

Arbequina: It is a small, brown olive, good for eating and for oil.

Arbosana: It is commonly grown for oil production.

Coratina: Olives are medium to large in size, with an oval shape and green color that turns black when ripe. They have high oil content (up to 25%) and produce oil with a robust flavor and aroma. Coratina olive oil is also rich in polyphenols, especially Oleacanthal, which has anti-inflammatory properties. The oil has a low acidity (< 0.5%).

Dolcy: Elongated black olives. Average size is 3–6 g. Percentage of oil is 15–18%. It is known for its sweetness or mild flavor.

Kalamata: It is considered as one of the best olive varieties. Fruits have large–dark purple color (Black). Fruit's shape is elongated with sharp point. Oil percentage is

15–20%. Fruit weight varies between 3 and 7 g. Seed represents 9–11% of the fruit weight. At harvest, fruits must be handpicked to avoid damage.

Koroneiki: A small olive, though difficult to cultivate, has a high yield of olive oil of exceptional quality. It has a high oil content (up to 28%) and produces excellent quality oil with a fruity, grassy, and peppery flavor. Its olive oil is rich in polyphenols, antioxidants that offer health benefits and stability to the oil. It also has a low acidity (< 0.8%).

Manzanilla: It is the most famous Spanish olive variety in Egypt. Fruits are almost rounded and have small to medium weights (4–6 g). Seed represents 10–12% of the fruit weight. Oil % varies from 16 to 20%. Fruits (Olives) start to ripen in September (Green Olives) until November (Black Olives). It has low oil content (up to 18%). Its olive oil is not rich in polyphenols but is high in Oleic acid, a mono-unsaturated fatty acid that can reduce cholesterol and improve cardiovascular health. The olive oil has a low acidity (< 0.8%).

Piqual: Egypt is a big producer of Picual olives. It has been mainly used as table olives for the past few years. It is small, rounded black olives. Sizes are medium 3-7 g, with an elliptical shape and a pointed oblique tip. Seeds are 12% of the fruit weight. Percentage of oil is 18–22%. It has a black color when ripe and has high oil content (up to 28%). Olive oil is also rich in polyphenols, especially Hydroxytyrosol, a powerful antioxidant that protects oil from oxidation and has a low acidity (< 0.4%).

2.4 Factors affecting olive production in Egypt

Several studies were conducted to identify the most important factors affecting the olive sector in Egypt. The most important internal factors affecting the strength of the olive sector are (1) the availability of arable land for growing olive trees, (2) the availability of good olive varieties, (3) climatic variability (weather fluctuations and strong winds), (4) high prices of production inputs and their unavailability in agricultural associations, (5) lack of agricultural mechanization lack of agricultural mechanization, (6) shortage of skilled labor and high wages, (7) adopting bad agricultural practices by producers, the most important is when and how much water to be used for irrigation (scheduling) during each stage of plant growth as well as the most efficient irrigation systems, (8) farmers' lack of awareness of the olive varieties that are suitable for the cultivation areas, (9) lack of a role for agricultural extension in guiding producers in production and manufacturing processes, and (10) management of olive crop and olive oil residues. While the most external factors were increasing the opportunities for exporting table olives and olive oil, and opening new external markets.

The studies suggested the following solutions for the previous problems: (1) expanding the arable areas in the desert through land reclamation projects, (2) breeding and selecting the appropriate olive varieties for the newly reclaimed areas, (3) cultivating windbreaks to face climate change, (4) provide agricultural associations with production inputs in sufficient quantities, at appropriate prices and in the required time, (5) educate the labor through competent organs in agricultural operations for the crop, (6) provide producers with the water requirements of the cultivated crops with the proper irrigation scheduling and highly efficient irrigation systems, (7) activate the role of agricultural extension in sensitization and orientation for the producers, and vv8) provide the best methods of utilizing the crop residues [25–27].

3. Egypt case study

The key principles associated with sustainability of olive production in Egypt including the use of the most efficient use of nonrenewable resources and on-farm resources are presented in this section.

3.1 Research related to irrigation systems, water and salinity management issues

3.1.1 Nassar

A field experiment was conducted for 3 years (2003–2006) in sandy soil to evaluate the effect of micro-irrigation systems (bubbler, mini-sprinkler, and trickle) on olive trees (*kalamata* cv.) [28]. The experimental site was arranged in a split plots design. The effect of different irrigation systems on following parameters was tested: morphology parameters, yield parameters, and soil parameters (moisture content and salt distribution in the soil profile). Results indicated that trickle-irrigation system was more effective than the other two systems. It increased the morphology parameters: trees height by 10 and 9.2%, stem diameter by 6 and 14%, shoots number/tree by 8 and 14%, leaves number/tree by 13 and 17%, shadow area by 6 and 13%, as well as increasing yield parameters: number of fruits by 16 and 24%, and olive yield (kg/tree) by 18 to 32% as compared with bubbler and micro-irrigation system, respectively. Results revealed also that root distribution under trickle irrigation showed more regularity than the other two types due to the placement and the number of irrigation distributors. Water productivity under trickle-irrigation system was 18 and 35% higher than bubbler and mini-sprinkler, respectively. It is recommended to use trickle irrigation as an efficient system for irrigating olive trees at Wadi El Natrown area in the west of the Nile Delta region. The main results are presented in (**Table 2**).

3.1.2 Attalla et al.

A 2-year field experiment was conducted during the two successive seasons of 2008 and 2009 on mature olive trees (*Manzanillo cultivar*) to test the effect of nine irrigation treatments applied from May to December on productivity and yield quality [29]. The tested treatments were: T1: rainfall (no irrigation), T2: 0.06 m once/month, T3: 0.08 m once/month, T4: 0.1 m once/month, T5: 0.12 m once/month, T6: 0.03 m twice/month, T7: 0.04 m twice/month, T8: 0.05 m twice/month, and T9: 0.06 m twice/month. The rainfall depths were 92.0 mm in 2008 and 115.0 mm in 2009. Irrigation water treatment (T9: 0.06 m twice/month) was more effective

Micro-irrigation system	Trees height (cm)	Stem diameter (mm)	Shoots number per tree	Leaves number per tree	Shadow area (cm^2)	Fruit weight (g)	Yield (t/ha)	WP (kg/m^3)
Trickle	155.26	6.87	4.13	423.57	125.52	4.76	43.67	25.71
Bubbler	141.22	6.44	3.78	366.06	117.52	4.64	35.91	21.14
Mini-sprinkler	128.54	5.93	3.55	350.93	108.95	4.3	30.04	17.69

Table 2.
Effect of irrigation systems on olive growth parameters, yield, and water productivity (WP).

in increasing the productivity and fruit quality of Manzanillo olive in both seasons compared to other treatments. Results also showed that leaf nitrogen, potassium, calcium, and iron contents were increased under such conditions. The main results are presented in (**Table 3**).

3.1.3 El-Taweel and Farag

The effect of three irrigation amounts, two drip irrigation methods, and two olive cultivars on growth parameters and yield of olive trees (*Olea europara* L.) were tested during the 2014 and 2015 seasons in a private orchard farm [30]. The tested variables were: (a) irrigation mounts: 10438 m^3/ha (100%), 8905 m^3/ha (85%), and 7355 m^3/ha (70%); (b) online surface and in-line subsurface drip irrigation methods, and (c) *Picual* and *Manzanillo* olive cultivars. A split-split experimental design with three replicates was used. The highest growth and yield were obtained from using 8905 m^3/ha irrigation amount as compared with other treatments. The 8905 m^3/ha applied amount was close to the estimated irrigation water by the FAO method. The online method recorded the highest olive and oil yields during the two seasons. Increasing the applied water decreased fruits' oil contents. The Manzanillo cultivar gave the highest yield while, Picual cultivar showed the highest oil content. The best-combined treatment of the tested factors was the 8905 m^3/ha + the online surface irrigation + Manzanillo cultivar. The highest water use efficiency was found in 7355 m^3/ha with subsurface irrigation combined with Manzanillo cultivar. The main results are presented in (**Table 4**).

3.1.4 Sary, Dalal, and Elsokkary

The *Picual, Aggizi, Mission, and Koroneiki* olive varieties were used to conduct a field experiment during the 2010/2011 growing season in calcareous soil. A split-split experimental design with three replicates was used to implement the study [31]. Full (FIW) and half irrigation water (HIW) of olive water requirement occupied the main plots. Evaporation pan was used to determine irrigation water. The four olive varieties occupied the sub-plots. Three NPK fertilizers (100, 75, and 50%) with effective

Treatment	Yield (kg/tree)		Fruit moisture (%)		Fruit oil content (%)	
	2008	2009	2008	2009	2008	2009
T1	21.31 e	1769 g	57.54 c	55.73 e	21.53 a	21.85 ab
T2	28.38 de	19.17 f	60.91 b	59.60 d	21.50 a	21.53 ab
T3	41.59 dc	20.93 e	60.86 b	61.58 c	21.38 b	21.30 ab
T4	52.14 ab	24.58 c	63.90 a	63.15 b	21.25 cd	21.00 ab
T5	57.12 a	26.39 b	64.39 a	6370 ab	21.23 cd	23.00 a
T6	34.95 d	19.65 ef	60.44 b	59.45 d	21.18 d	21.75 ab
T7	38.19 d	22.56 d	6164 b	61.65 c	21.23 cd	21.30 ab
T8	47.12 bc	25.91 b	64.13 a	63.38 b	21.28 c	21.13 ab
T9	58.59 a	27.83 a	63.44 a	64.40 a	21.20 cd	20.55 b

5% probability level ($p < 0.05$).

Table 3.
Fruit yield, fruit moisture, and oil content are affected by irrigation treatments.

Treatments		Yield/tree (kg) (first season)		Yield/tree (kg) (second season)		Oil content (%) (first season)		Oil content (%) (second season)	
Irrigation level	Irrigation method	Picual	Manzanillo	Picual	Manzanillo	Picual	Manzanillo	Picual	Manzanillo
10,438 m^3/ha	In	22.1	31.7	8.4	12.0	32.0	31.5	32.4	32.0
	On	2.9	21.7	1.1	6.2	30.5	29.8	30.7	30.0
8905 m^3/ha	In	23.3	33.9	9.8	12.9	36.3	33.6	36.8	34.1
	On	19.7	23.3	7.4	8.6	32.5	31.4	33.0	31.9
7355 m^3/ha	In	18.8	25.9	7.1	9.8	37.8	35.5	38.3	36.0
	On	21.1	30.3	8.0	11.5	35.2	33.2	35.7	33.6

Table 4.
Effect of tested treatments on yield and fruit oil contents in the two seasons.

microorganisms (EM) or humic acid (HA) occupied the sub-sub-plots. There were significant effects of FIW or HIW, 75% NPK + EM, or 50% NPK + EM on fruit weight of Picual and Aggizi. All treatments showed significant effects on increasing fruit weight of Mission. The Koroneiki variety showed no significant response to the tested treatments. Oil contents of Picual, Aggizi, Mission, and Koronieki fruits showed a significant response to 100% NPK, 75% NPK + EM, or 75% NPK + HA treatments under a full irrigation regime. The highest significant values of leaves' chlorophyll contents in all olive varieties were obtained by full irrigation with 75% NPK + EM and 75% NPK + HA. Leaves' contents of N, P and K in the four olive varieties significantly increased as a result of applying all treatments under FIW or HIW regimes. All treatments significantly increased the levels of total N, available P, and available K in the soil, as well as soil respiration with FIW or HIW regimes. The main results are illustrated in (**Figure 3**).

3.1.5 Ali et al.

A study was conducted during two successive seasons (2019 and 2020) to evaluate the performance of seven fruitful olive cultivars (*Olea europaea L.*): *Aggizi Shami*, *Dolce*, *Picual*, *Manzanillo*, *Coratina*, *Koroneiki*, and *Arbequina* in saline calcareous soils [32]. The vegetative growth, flowering, yield, fruit characteristics, and oil

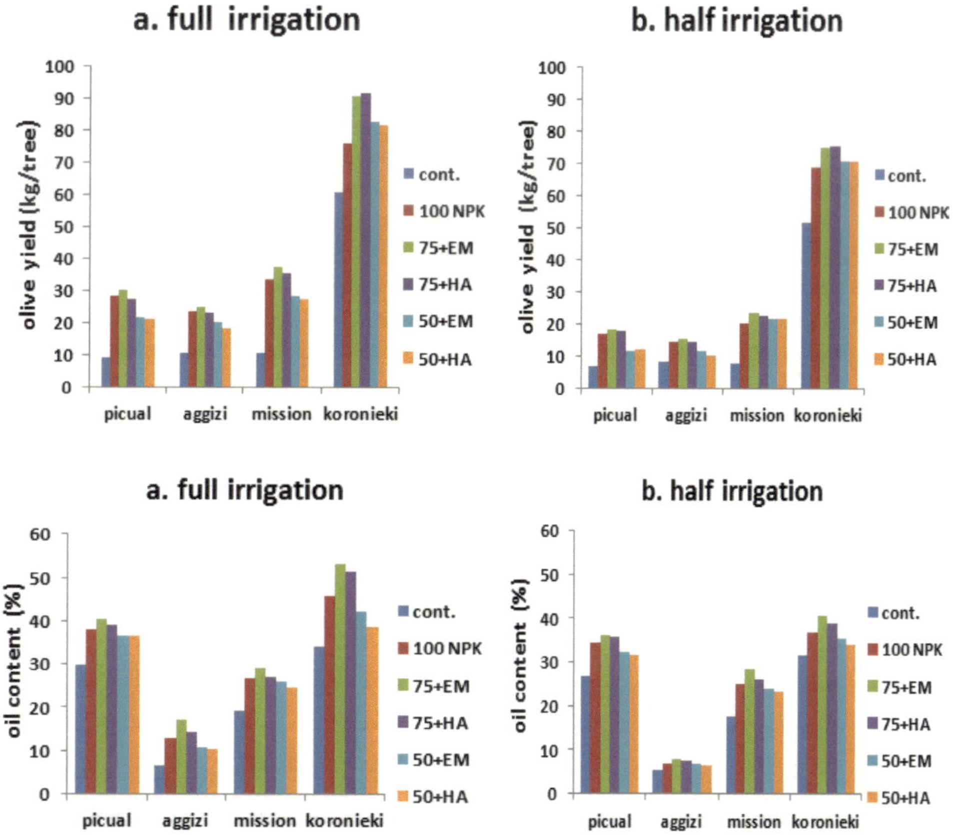

Figure 3.
Effect of tested treatments on fruit olive and oil yields.

content parameters, as well as salinity stress tolerance of the seven varieties, were tested. Coratina and Koroneiki cultivars produced the longest canopy volume, while Arbequina, followed by Dolce cultivars, had the smallest ones. Ideal flower percentages differed significantly among the studied cultivars and seasons. The highest yields were recorded for the Aggizi Shami variety, while the Arbequina variety recorded the lowest yield weight (kg/tree) in both seasons. Koroneiki and Coratina varieties recorded the highest oil (%), while Aggizi Shami and Dolce recorded the lowest percentages. Picual and Arbequina cultivars had the highest leaf proline contents, while Dolce cultivars had the lowest value. Results concluded that Coratina, Koroneiki, Picual, and Aggizi Shami varieties can be recommended under the studied conditions. The main results are presented in (**Table 5**).

3.1.6 Alowaiesh et al.

A field experiment was conducted during the 2020 and 2021 growing seasons to evaluate the effect of medium crop water requirements (75% ETc), severe water stress (50% ETc), and full (100% ETc) on growth and physio-chemical parameters of *Manzanillo*, *Eggizi-Shami*, and *Tofahi* olive cultivars [33]. Gene expressions including dehydration-responsive element binding (DREB), dehydrin (DHN), and catalase (CAT) genes in olive cultivars were texted under the three irrigation regimes. Considerable effect of irrigation level on all studied parameters was recorded. Medium and severe stress treatments caused a gradual reduction in nitrogen, phosphorus, and potassium contents, relative water content, root and shoot length, root and leaf numbers, branch count, and leaf area in the tested varieties. Water stress treatments increased proline contents. Eggizi-Shami variety proved to be superior in all tested parameters. The real-time quantitative PCR (RT-qPCR) analysis displayed significant changes in gene expression of the three genes (DHN, DREB, and CAT) as related to water stress response. The RT-qPCR analysis under the stress conditions showed that Eggizi-Shami exhibited higher expression than Tofahi and Manzanillo cultivars. Based on the morphological and physiological parameters with gene expression analysis under water stress conditions, results indicated that related genes can offer highly validated information about selecting drought-tolerant olive cultivars. The main results are presented in (**Table 6**).

Olive variety	Relative water content (%)	Total chlorophyll (mg/g fw)	Proline (μg/g fw)	Oil% (fruit dry weight)	Yield (kg/tree)
Aggizi Shami	79.435	2.975	41.21	27.4	33.26
Dolce	72.02	2.815	37.42	32.61	27.79
Picual	83.33	3.250	43.29	38.42	25.3
Manzanillo	77.07	2.920	39.99	36.63	26.08
Coratina	75.74	2.885	39.31	45.85	24.38
Koroneiki	79.25	2.960	41.11	46.61	20.68
Arbequina	80.63	3.060	41.84	38.68	16.6

Table 5.
Effect of salinity stress on average relative water content, total chlorophyll, proline content, oil%, and fruit yield of the tested olive cultivars.

Factor	Proline (mg/g FW)		N (%)		P (%)		K (%)		No. of roots		No. of branches		Leaf area (cm^2)		No. of leaves	
Season	**first**	**second**	**first**	**second**	**first**	**second**	**first**	**second**	**first**	**second**	**first**	**second**	**first**	**second**	**first**	**second**
Irrigation level:																
100% ETc	19.9c	20.1b	1.98a	2.04a	0.20a	0.2 a	1.06a	1.11a	24.3a	26.3a	14.9a	17.8a	4.16b	4.23a	84.67a	88.56 a
75% ETc	20.6a	20.9b	1.87b	1.94b	0.19b	0.18 b	0.98b	1.07a	16.0 b	21.1b	11.67b	13.9b	4.13ab	4.1ab	58.22b	63.44b
50% ETc	21.9a	22.2a	1.8b	1.87b	0.18b	0.18 b	0.83b	0.90b	6.40 c	10.0c	9.47c	11.8c	4.07b	4.07b	42.89c	46.33c
Cultivar:																
Manzanillo	22.4a	22.6a	1.83 b	1.90 b	0.19b	0.18 c	0.95 b	1.04 b	14.4 b	18.1b	11.1 b	13.7 b	4.09 b	4.13 b	58.9 c	62.7 c
Eggizi-Shami	23.0a	23.2a	2.03 a	2.11 a	0.20 a	0.190 a	1.06 a	1.17 a	15.1 b	19.1ab	13.3 a	15.6 a	4.15 a	4.15 a	66.2 a	69.8 a
Tofahi	16.8b	17.4b	1.78 b	1.84 b	0.19b	0.19 b	0.86 b	0.87 c	17.2 a	20.2 a	11.7 b	14.2 ab	4.12 a	4.13 b	60.7 b	65.9 b

5% probability level ($p < 0.05$).

Table 6.
Impact of different irrigation levels on the physio-chemical constituents and growth characteristics of the three olive varieties in the 2020 and 2021 seasons.

3.1.7 Taha and Khalifa

A two-season field experiment was conducted on drip-rigidated *Coratina cultivar* olive oil orchard in a private farm during the 2020 and 2021 seasons [34]. The study was conducted to evaluate the effect of different levels of reference evapotranspiration (ETo) (e.g., 120, 100, 80, and 60% ETo), and farmer practice on amounts of applied irrigation water (AIW), consumptive use (WCU), olive fruit and oil yields and some fruit quality parameters, water use efficiency (WUE), water productivity (WP), electric energy consumed, farm income, and benefit-cost ratio (BCR). Also, to develop a local olive coefficient (Kc) and yield response factor (Ky) under experimental conditions. A randomized complete blocks design with four replicates was used. Results revealed that average reference evapotranspiration values varied between 36 mm/month in January and 166.8 mm/month in July. Average AIW values were 9953, 8484, 6971, 5480, and 17,488 for the 120, 100, 80, and 60% ETo treatments and farmer practice, respectively. Olive fruit and oil yields and quality parameters were significantly affected by the tested irrigation treatments. The highest fruit yields of 11.31 and 9.34 t/ha were recorded for the 120% ETo treatment, while the lowest values of 7.048 and 5.75 t/ha were recorded for the 60% ETo treatment in the first and second seasons, respectively. Average oil yields were 1.49, 1.435, 1.365, 1.15, and 1.18 t/ha for the 120, 100, 80, and 60% ETo treatments and farmer practice, respectively. The Kc value of 0.86 and Ky value of 0.83 were obtained for the Coratina olive variety for the 120% ETo treatment. Average consumed energy values were 43.2% for the 120% ETo and 68.9% for the 60% ETo treatments less than farmer practice. Average net income values for the 120, 100, 80, and 60% ETo treatments were 33, 28, 23, and 3%, respectively, higher than the net income of farmer practice. The BCR values for the same respective treatments were (15.9, 16.7), (15.1, 17.7), (14.8, 18.7), (13.3, 17.5), and (10, 9.2) for olive fruit and oil. The highest 2-year average WUE of 1.71 kg fruits and 0.33 kg oil/m^3 consumed water were recorded for the 60% ETo treatment, while the lowest values of 0.71 kg fruits and 0.08 kg oil/m^3 consumed water were obtained from farmer practice. The WP values of the same treatments were 1.29 kg fruits and 0.25 kg oil/m^3 applied water and 0.58 kg fruits and 0.07 kg oil/m^3 applied water. Based on the net income, the highest WP value of 18.93 LE/m^3 was recorded for the 60% ETo, while the lowest value of 5.73 LE/m^3 was obtained from the farmer practice. Producing oil from olive trees (Coratina var.) in sandy soils can be achieved by applying amounts of water equal to 80% or 60% ETo. It is also recommended to convey the achieved results to the farmers in the region and similar areas by the extension authorities. Some of the obtained relations are illustrated in (**Figures 4–7**).

3.1.8 Khalifa and Taha

A field experiment was conducted on drip-irrigated table olive trees (*Aggizi cultivar*) in a private farm during the 2020 and 2021 seasons to evaluate the effect of five irrigation treatments (120, 100, 80, and 60% ETo, and farmer practice) on amounts of applied irrigation water (AIW), consumptive use (WCU), olive fruit yield, some physical parameters, water use efficiency (WUE), water productivity (WP), electric energy used, and farm income [35]. Also, to develop a local table olive coefficient (Kc) and yield response factor (Ky). The experiment was laid out in a randomized complete blocks design with four replicates. Results revealed that the average reference evapotranspiration values varied between 1.16 mm day^{-1} in January and 5.35 mm day^{-1} in July. Results indicated also that, the 2–year average

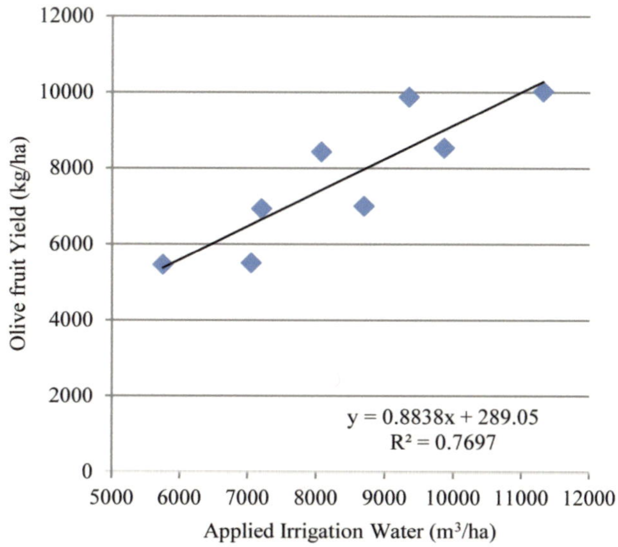

Figure 4.
Linear relation between applied water and olive fruit yield.

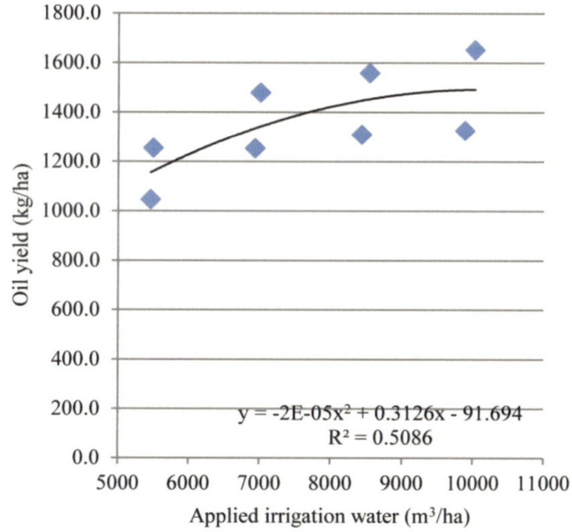

Figure 5.
The second-order equation for the relation between oil yield and applied water.

AIW values were 9952 (14.94 m^3 $tree^{-1}$ $yr.^{-1}$), 8484 (12.74 m^3 $tree^{-1}$ $yr.^{-1}$), 6971 (10.71 m^3 $tree^{-1}$ $yr.^{-1}$), 5480 (8.23 m^3 $tree^{-1}$ $yr.^{-1}$), and 14,747 m^3 ha^{-1} (21.61 m^3 $tree^{-1}$ $yr.^{-1}$) for the 120, 100, 80, and 60% ETo treatments and farmer practice, respectively. The WCU values varied from 7817 m^3 ha^{-1} (11.7 m^3 $tree^{-1}$ $yr.^{-1}$) to 3359 m^3 ha^{-1} (5.0 m^3 $tree^{-1}$ $yr.^{-1}$) for the 120 and 60% ETo treatments and 11,437 m^3 ha^{-1} (17.2 m^3 $tree^{-1}$ $yr.^{-1}$) for farmer practice. Olive fruit yields and physical parameters were significantly affected by the tested treatments. The highest fruit yield of 17.08 and 14.6 t ha^{-1} was recorded for the 120% ETo treatment, while the lowest yield of 11.12 and 9.66 t ha^{-1} was obtained from the 60% ETo treatment

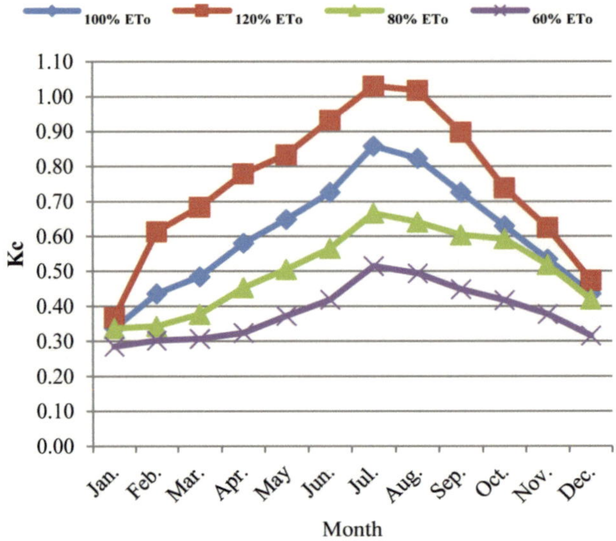

Figure 6.
Crop coefficients of the tested irrigation treatments.

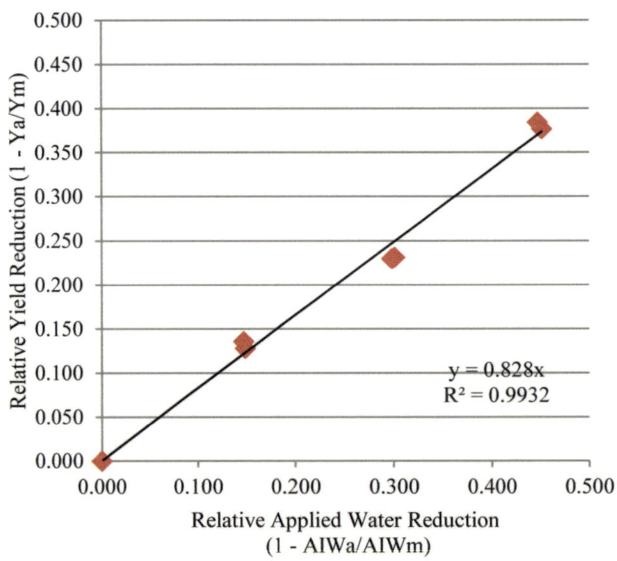

Figure 7.
Yield response factor for Coratina var.

in the first and second seasons, respectively. The highest WUE of 3.09 kg m^{-3} consumed water was recorded for the 60% ETo treatment, while the lowest value of 1.0 kg m^{-3} consumed water was obtained from farmer practice. The WP values of the same treatments were 1.9 and 0.8 kg fruits m^{-3} applied water. For the Aggizi olive cultivar, a Kc value of 0.77 and a Ky value of 0.836 were obtained. Average consumed energy values varied from 32.5% for the 120% ETo to 62% for the 60% ETo treatments less than farmer practice. Average net income values for the 120%, 100%, and 80% ETo treatments were 47.6, 18.1, and 8.2% higher than farmer practice. Irrigating olive trees (Aggizi cultivar) in sandy soils with amounts of water equal to 120% ETo will save 32.5% of applied irrigation water and energy used, achieve WUE of 2.02 kg fruits m^{-3} of consumed water and WP of 1.6 kg fruits and 17.98 LE m^{-3} of applied water, as well as 47.6% higher net income as compared with farmer practice. In case of water shortage, it is recommended to apply the 80% ETo treatment, since it will save water and energy and achieve higher net income and water productivity than farmer practice. It is also recommended to convey the achieved results to the farmers in the region and similar areas by the extension authorities. Some of the obtained relations are illustrated in (Figures **8–10**).

3.2 Research related to nutrients and microorganisms issues

3.2.1 El-Fouly et al.

An experiment was carried out at a private farm located at km 48 of Alexandria road in the northwest of Egypt, between January 2011 and December 2012 on 10-year-old olive trees cv [36]. *Koroneiki* grown on a loamy sand soil to evaluate

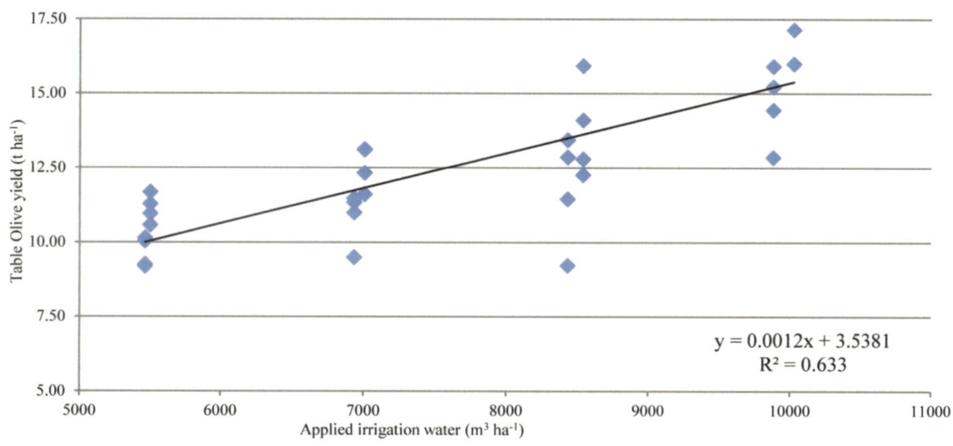

Figure 8.
Linear relation between applied irrigation water and table olive yield.

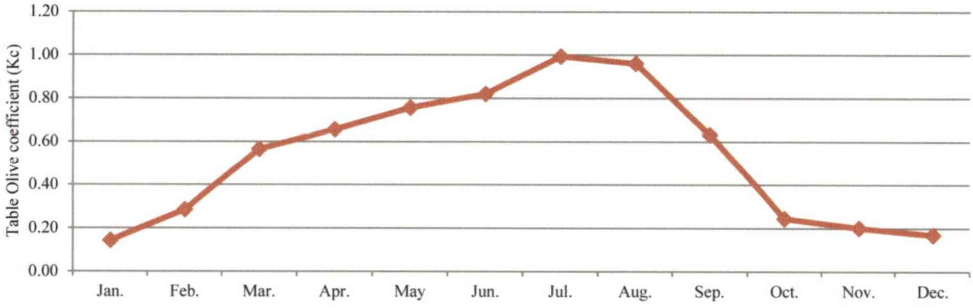

Figure 9.
Crop coefficient of the 120% ETo treatment.

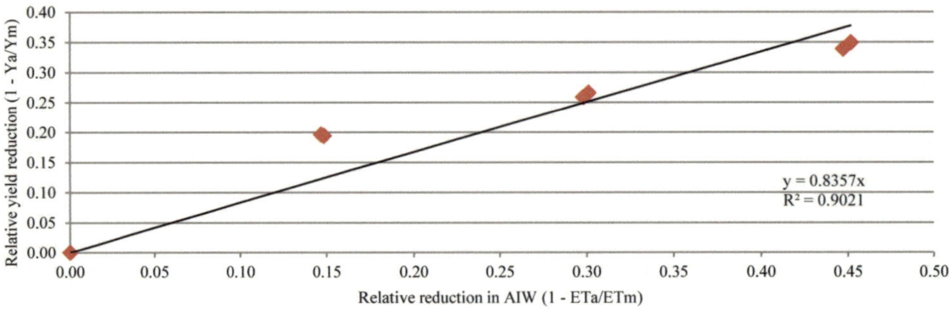

Figure 10.
Yield response factor for Aggizi var.

nutrient status and to calculate the amount of nutrients annually removed by the olive tree. The distance between trees was 6 × 6 meters (278 tree ha^{-1}). All cultural practices were followed and trees were subjected to moderate pruning every year. Nutrient concentrations in the flesh and pit were determined and used to quantify the nutrients removed by fruit. The amounts of nutrients removed annually during pruning were calculated based on the dry matter and nutrient concentration in the different tissues. The Costat statistical package was used to calculate means, maximum, minimum, and standard deviations (SD), with 12 replicates. Results showed that leaves have sufficient contents of N, P, K, Ca, Na, Mn, Zn, Cu, and B, while the level of both Fe and Mg was high. Nutrients removed annually/tree were 265.24 g N, 37.93 g P, 353.93 g K, 122.67 g Ca, 76.94 g Mg, 74.78 g Na, 7.288 g Fe, 0.773 g Mn, 0.514 g Zn, 0.213 g Cu, and 0.663 g when the yield was 77.33 kg/tree. The results obtained proved to be helpful in calculating fertilizer recommendations. Average nutrient removal from different part of olive tree is presented in (**Table** 7).

Nutrient	Fruit (g/tree)			Pruned material (g/tree)			Total (g) (nutrient/ yr./tree)
	Flesh	Pit	Total	Branches	Leaves	Total	
N	140.70	96	237.00	10.21	18.03	28.24	265.24
P	21.61	7.89	29.50	6.26	2.17	8.43	37.93
K	290.00	19.73	309.70	27.08	17.15	44.23	353.93
Ca	15.23	30.77	46.00	36.41	40.26	76.67	122.67
Mg	26.10	26.83	52.90	19.09	4.95	24.04	76.94
Na	44.23	15.78	60.00	11.10	3.68	14.78	74.78
Fe	3.44	1.66	5.10	1.54	0.65	12.19	7.29
Mn	0.19	0.14	0.33	0.34	0.11	0.45	0.77
Zn	0.22	0.13	0.35	0.12	0.04	0.16	0.51
Cu	0.12	0.06	0.17	0.03	0.01	0.04	0.21
B	0.45	0.07	0.52	0.12	0.03	0.14	0.66

Table 7.
Nutrients removed from fruits, pruned material and total nutrient/year/per Koroneiki olive tree (g) on dry weight bases.

3.2.2 Abd-Alhamid et al.

A field experiment was conducted in a private orchard farm (latitude 32.26° N, longitude 30.33° E, and altitude 10 m above mean sea level), Egypt to study the effect of different levels of mineral nitrogen fertilization with bio-fertilizer on fruit set, yield, fruit quality, oil properties and total microbial count [37]. The experiment was carried out on 15-year-old olive trees (cv. *Manzanillo*), planted 5 × 5 m apart (400 trees ha^{-1}) in sandy soil under a drip irrigation system. The olive trees yearly nutrient requirements were equal to 1000 g N/tree/year. Ammonium sulfate (20.6%N) as a source of N-fertilizer was used. Also, 1.75 kg/tree of super phosphate (15.5% P_2O_5) and 1.50 kg/tree of potassium sulfate (48% K_2O) were used as soil applications in two equal doses. Also, microbial cultures and bio-fertilizers were inoculated. The biofertilizer used consisted of liquid cultures of three bacteria: *Azotobacter chrococcum, Bacillus megaterium, and Bacillus circulans*. Cultures were mixed on site and were applied at the rate 2 liters/tree. This treatment was added three times during the season. The tested treatments were: T_1: control, 100% mineral N (1000 g N/tree); T_2: 75% mineral N (750 g N/tree + biofertilizer, 2 liters/tree); T_3: 50% mineral N (500 g N/tree + biofertilizer, 2 liters/tree); and T_4: 25% mineral N (250 g N/tree + biofertilizer, 2 liters/tree). The results indicated that the highest yield of 52.95 kg/tree (21.18 t ha^{-1}) was recorded for the T_2 treatment. It was also the most effective treatment in enhancing fruit weight, volume, length, fruit oil content, oil quality, and total microbial count. The main results are presented in (**Table 8**).

3.2.3 Abd El-Migeed et al.

This work was carried out in an olive farm located in Nekhl Center, Sinai, on 5-year-old olive trees to evaluate the effects of foliar application of different algae concentrations on growth, yield, and fruit quality of olive trees cv. *Koroneiki* is grown in sandy calcareous soil under saline conditions [38]. The trees were almost uniformly spaced 3 × 5 m irrigated by drip system. The tested treatments were: T1, control 0.0% algae extract (AE); T2, 0.05% algae extract; T3, 0.075% algae extract; and T4, 0.1% algae extract. Results in two growing seasons showed that, all algae extract treatments improved the most morphological characters and yield components compared with the control. Results also revealed that increasing AE increased vegetative growth and productivity parameters. The study recommended spraying 0.1% AE three times, i.e., at growth start, at flowering, and after the final fruit set, to improve the growth, yield, fruit quality, and oil content of Koroneiki olive. (**Table 9**) shows the main results of the study.

3.2.4 Abd El-Razek et al.

A 2-year study was carried out on '*Kalamata*' olive trees (*Olea europaea* L.) in a private orchard farm. Drip irrigation system was used in the farm [39]. Four soil application treatments were used: (T1) control (water only); (T2) 10 kg chicken manure was added in both side of the tree, (T3) 10 kg chicken manure +100 cm^3 humic acid (Actosol)/tree, (T4) 10 kg chicken manure + bio-humic (containing 100 cm^3 Actosol +150 cm^3 of *Azotobacter chroococcum*, *Bacillus megaterium*, and *Bacillus circulans* in equal doses/tree). Humic acid and bio-humic were added three times during March (full bloom), May (starting fruit set stage), and at the third stage of fruit development (70% of final fruit size) to study the effect of humic acid and

Treatments	Yield (kg/tree)				Fruit weight (g)				Flesh oil content (%)			
	2014	2013	2012	Mean	2014	2013	2012	Mean	2014	2013	2012	Mean
B + MNF100%	68.07	26.38	57.37	50.61	5.22	8.6	6.09	6.64	40.27	33.33	28.44	34.01
B + MNF75%	71.67	28.13	59.97	53.26	5.73	8.96	6.87	7.19	41.13	32.33	29.8	34.42
B + MNF50%	64.34	26.37	59.1	49.94	5.19	8.57	5.65	6.47	38.67	32.06	28.08	32.94
B + MNF25%	51.47	22.67	52.33	42.16	4.44	7.93	5.38	5.92	35.89	30.89	27.44	31.41

Table 8.
Effect of tested treatments on olive yield, fruit weight, and flesh oil content.

Parameter treatment	Yield (kg/tree)	Fruit weight (g)	Flesh weight (g)	Seed weight (g)	Flesh/frit ratio	Flesh oil (%)
T1: AE (0) control	16.43	1.64	1.19	0.46	0.72	16.75
T2: AE (50 cm^3/100 L.W.)	17.73	1.72	1.18	0.55	0.68	19.08
T3: AE (75 cm^3/100 L.W.)	20.83	1.78	1.26	0.52	0.71	21.26
T4: AE (100 cm^3/100 L.W.)	24.50	1.80	1.31	0.49	0.73	22.20
AE = Algae extract	L.W. = Liters of water					

Table 9.
Effect of sea algae extracts on yield and fruit quality of Koroneiki olive trees.

bio-humic on the yield and fruit quality of Kalamata olive trees. Results showed that all treatments increased the nutrient status (N, P, K, Ca, Mg, Fe, Zn, Mn, and Cu) of the leaves, yield (kg/tree), and fruit quality parameters, as well as fruit moisture content (%) and oil (%) in fresh and dry weights compared to the control. The bio-humic treatment (T4) is recommended, since it had the highest values of the parameters in comparison with other treatments. Results indicated that improving yield and fruit quality are related to the positive effect of organic matter alone or in combination with humic or bio-humic in increasing the cation exchange capacity of the soil, reducing soil pH, enhancing the root development, increasing the root/shoot ratio, and production of root hairs of olive trees with direct effect on increasing the active uptake for most nutrients in the soil. Bio-humic contains three bacteria that are considered as plant growth-promoting rhizobacteria (PGPR) and play a great role in providing trees with NPK as bio-fertilizers. Bio-humic had positive effects reflected on improving the yield and fruit quality of Kalamata olive trees. The application of organic manure in combination with bio-humic had great effects on improving the yield and fruit quality of Kalamata olive trees compared with using organic manure alone or organic manure combined with humic acid. The main results are summarized in (**Table 10**).

3.3 Research related to the performance of different olive cultivars

3.3.1 Saad El-Din, Ikram et al.

A three-year study was conducted to evaluate the performance of progenies from crosses between cvs. (*Chemlali* × *Aggizi*), (Aggizi × *Kalamata*), and (Aggizi × *Koronaiki*) [40]. The progenies were analyzed for the (shoot length, shoot thickness, number of nodes/shoot, internodes length (cm)), the leaves (average number of leaves/shoot, leaf surface area, leaf shape), flowering (flowering time, the length of inflorescence, number of total flowers/inflorescence, number of perfect flowers/inflorescence, number of staminate flowers/inflorescence, sex ratio, and fruit set/m), fruiting (production, fruit shape, fruit weight, stone weight, flesh weight, flesh/stone, moisture and oil content (oil percent in fresh weight and oil percent in dry weight) and rooting ability characters of olive tree. From the obtained results, it could be concluded that progenies no. 24, 26, 27, and 31 can be selected for table olive; progenies no 7, and 10 for oil; and progenies no. 33, 38, and 39 can be selected for dual purposes. A summary of recorded results is given in (**Table 11**).

Treatment	Yield (kg/tree)			Fruit weight (g)			Oil (%) – dry weight			Oil (%) – fresh weight		
	2016	2017	Mean	2016	2017	Mean	2016	2017	Mean	2016	2017	Mean
T1	43.33	48.31	45.82	5.15	5.44	5.30	35.10	36.20	35.65	16.36	18.90	17.63
T2	70.67	60.27	65.47	5.66	6.16	5.91	39.65	38.80	39.23	22.26	22.10	22.18
T3	81.00	75.43	78.22	6.22	7.23	6.73	41.03	40.33	40.68	26.08	25.77	25.93
T4	89.67	83.65	86.66	6.52	7.74	7.13	45.00	43.07	44.04	31.56	32.11	31.84

T1: control; T2: chicken manure; T3: chicken manure + Humic Acid; T4: chicken manure + Bio-Humic.

Table 10.
Effect of tested treatments on yield, fruit weight, oil (%) on dry and fresh weights.

Selections	♀X♂	Productivity	Constant productivity	Fruit weight	Flesh/ stone	Oil % (dry weight)
2	Chemlali × Aggizi	43.92	Alternate	Very low	Medium	Medium
3	Chemlali × Aggizi	43.67	Alternate	Very low	Medium	Medium
4	Chemlali × Aggizi	40.92	Alternate	Very low	Medium	Medium
6	Chemlali × Aggizi	41.83	Alternate	Very low	Medium	Medium
7	Chemlali × Aggizi	39	Alternate	Very low	Medium	High
10	Chemlali × Aggizi	43.75	Alternate	Very low	Medium	High
16	Aggizi × Kalamata	36.08	Alternate	High	Medium	Low
18	Aggizi × Kalamata	36	Alternate	High	Medium	Low
19	Aggizi × Kalamata	39.42	Alternate	High	Low	Low
20	Aggizi × Kalamata	39.67	Alternate	High	Medium	Low
24	Aggizi × Kalamata	43	Alternate	High	Medium	Low
26	Aggizi × Kalamata	47.33	Alternate	High	Medium	Medium
27	Aggizi × Kalamata	44.08	Constant	High	Medium	Medium
28	Aggizi × Kalamata	38.17	Alternate	High	Medium	Medium
31	Aggizi × Kalamata	32.67	Alternate	High	High	Medium
33	Aggizi × Koronaiki	47.33	Constant	Medium	High	High
38	Aggizi × Koronaiki	49.92	Constant	Medium	High	High
39	Aggizi ×Koronaiki	48.67	Constant	Medium	High	High
Aggizi		41.67	Alternate	High	Medium	Very Low
Kalamata		35.33	Alternate	High	Medium	High
Chemlali		51.5	Constant	Very Low	Low	High
Koroneiki		40.08	Alternate	Very low	Medium	High

Table 11.
Summary of the recorded results.

3.3.2 Hegazi

The performance of 12 olive cultivars introduced to Egypt in 1984 (*Thrombolia, Strogylia, Villalonga, Cerasicola, Dermlali, Leccio Dela Corna, Roseciola, Boutellian, Ouslati, Mouslati, Enduri, and Tansh*) were investigated under Egyptian conditions (Giza Governorate) for two seasons 2008 and 2009 [41]. The vegetative characteristics, floral biology, fruit set, fruit characteristics and oil percentage of these cultivars were evaluated. In the first season, results showed that dates of full bloom (FB) for bud break of Tansh, Roseciola, and Leccio Dela Corna cultivars were significantly late compared to other cvs., while the earliest date was recorded for Mouslati cv. Growing degree days (GDD) for full bloom was significantly higher in Mouslati cv. compared to other cvs., while it was significantly lower in Tansh cv. Highest leaf density was recorded for Boutellian cv., while it was the lowest in Ouslati and Villalonga cvs. No significant differences between average leaf area of Villalonga, Mouslati, Leccio Dela Corna, Ouslati, Cerasicola, Roseciola, Boutellian, and Strogylia cvs, while it was the lowest in Enduri, Tansh, Dermlali and Thrombolia cvs. As for flowering density, Enduri, Ouslati, Cerasicola, and Villalonga cvs. Were significantly higher than the other cvs. Number of perfect flowers/inflorescence was significantly higher in Dermlali cv. and it was lowest in Leccio Dela Corna and Enduri cvs. The highest fruit sets were recorded for Tansh, Ouslati, and Leccio Dela Corna, while it was the lowest in Thrombolia cv. As for fruit length, diameter, weight, and flesh weight parameters, the Strogylia and Boutellian cvs. Were the highest, while the lowest was observed in Ouslati cv. Oil percentage for Leccio Dela Corna, Boutellian, and Mouslati was the highest, while the lowest oil% was observed in Cerasicola and Strogylia cvs. The study concluded that Leccio Dela Corna, Boutellian, and Mouslati olives were superior in oil content, while Strogylia and Boutellian were superior in the other fruit characteristics. The main results are presented in (**Table 12**).

3.3.3 Fayek et al.

Three local olive clones '*Sewia*', '*Maraki*', and '*E52*' were compared with two international cultivars '*Coratina*' and '*Koroneiki*', during two seasons. Flowering time varied according to cultivars and growing season [42]. 'Coratina', 'Maraki', and 'Sewia' were earlier in flowering than 'Koroneiki' but were late for E52. The 'Coratina var.' had longer inflorescence (>3.5 cm) than the other cultivars, which were medium (2.5–3.5 cm). 'Maraki var.' recorded the highest number of flowers per inflorescence (>25), while the lowest number (<18) was recorded for 'Koroneiki'. Perfect flower percentage was affected by cultivars and the growing season. It was the highest in 'Coratina' (>90%) and the lowest in 'Sewia' (75%). 'Sewia var.' recorded the highest fertility of pollen, and the lowest was in 'E52'. All of the studied cultivars were self-incompatible. Under open pollination conditions, 'Coratina' produced the highest yield in the two growing seasons, while the 'E52' was the alternate bearing cultivar. Fruit and stone weights were lowest for 'Koroneiki var.', medium in 'E52', high in 'Coratina' and 'Maraki', and very high in 'Sewia'. Fruit shape was elongated in 'Koroneiki', 'ovoid-elongated' in 'E52' and 'Coratina', and ovoid in 'Sewia'. Stone shape was elongated in 'Koroneiki', 'E52', and 'Coratina'; ovoid in 'Maraki' and 'elliptic' in 'Sewia'. Oil content (%) was highest in 'Maraki' and 'E52' followed by 'Coratina', 'Sewia', then 'Koroneiki' in decreasing order. The total unsaturated/saturated fatty acids ratio was highest in

Cultivars	Fruit weight (g)		Seed weight (g)		Oil (%)	
	2008	2009	2008	2009	2008	2009
Thrombolia	6.29	6.47	0.80	0.85	29.33	29.33
Strogylia	8.59	8.80	0.78	1.02	27.33	27.33
Villalonga	5.87	6.00	0.97	0.93	30.33	31.33
Cerasicola	3.75	3.89	1.04	1.11	27.00	27.33
Dermlali	4.12	4.17	0.88	0.89	33.33	34.00
Leccio Dela Corna	2.54	2.53	0.43	0.50	35.67	36.00
Roseciola	3.58	3.70	0.72	0.76	30.33	31.67
Boutellian	8.90	8.73	0.65	0.69	35.33	36.00
Ouslati	1.85	2.13	0.23	0.67	28.33	29.00
Mouslati	1.88	1.82	0.47	0.39	35.33	36.00
Enduri	2.38	2.52	0.35	0.47	33.67	34.67
Tansh	6.77	6.87	0.89	0.91	29.00	29.67
LSD (0.05)	**0.523**	**0.294**	**0.337**	**0.285**	**1.151**	**1.265**

Table 12.
Fruit weight, seed weight, and oil content (%) of the evaluated olive cultivars.

'Maraki' followed by 'Sewia' then 'Coratina', but the lowest ratio was found in 'E52' and 'Koroneiki'. Based on RAPD and ISSR-PCR genetic markers, genetic similarity was found between 'Sweia' and 'Maraki' and the least between 'Coratina' and 'Koroneiki'. The main findings are presented in (**Table 13**).

3.3.4 Omran

The performance of 12 local and national genotypes (20 years old) was evaluated and their traits were compared with those of their parental cultivars for three seasons [43]. The selected cultivars were: [*Aggizi* and *Toffahi* (the main table olive cvs., 'Egypt'); *Chemlali* (olive oil cv., 'Tunisia'); *Arbequin* (oil cv.); and *Manzanillo* (dual cv. 'Spain'), as well as *Kalamata* (table olive cv. 'Greece')]. The genotypes and parental cultivars were analyzed for different traits including shoot, leaf, floral, fruits, fruit moisture, oil content, and rooting ability. The greatest values of most studied traits were significantly found in genotype 14. Toffahi cv. (third season) recorded the earliest date of flowering, while Arbequin cv. was the latest one (first season). The other genotypes and parental cultivars showed the greatest total number of flowers and perfect flowers/inflorescence ratio in genotypes 13 and 14, the highest percent of perfect flowers, and the minimum of staminate (male) flowers in genotype 14. The highest fruit set % was obtained from the 14 and 79 genotypes. A group of genotypes derived from (Arbequin x Aggizi) recorded the highest crop yields and are classified as high oil content. The tested genotypes can be classified as follows: genotypes 15 and 68 (table olives), genotypes 53 and 69 (dual use), and genotypes 14, 79, 85, and 86 (oil production). The Arbequin cultivar was the highest in the ability to rooting%. The main results are presented in the following (**Table 14**).

3.3.5 Arafat et al.

The six olive varieties *66, 69, Arbosana, Oleana, Koroneiki, and Arbequina*, which differ in their oil composition, were studied under local conditions [44]. Some characteristics of the extracted oils were evaluated. Results revealed that (1) there were significant differences between the oils of the studied varieties, and (2) the majority of the studied analytical variables were significantly affected by the interaction between the variety and the environment. The study concluded that Koroneiki and variety no. 69 can be cultivated by highly intensive culture to produce olive oil with high quality. Results in (**Table 15**) show the chemical composition (on dry basis) and pigments of chlorophyll and carotene (mg/kg) of the six investigated olive varieties.

Cultivars	Yield (kg/tree)		Oil content (%)		Oil yield (kg/tree)	
	2011	2012	2011	2012	2011	2012
Sewia	45.00 A	46.33 B	46.22 C	50.86 B	8.83 B	10.24 B
Maraki	41.67 AB	55.73 A	52.82 A	53.45 A	9.65 AB	13.39 A
E52	36.67 B	7.50 C	51.23 A	54.54 A	8.78 B	5.80 D
Koroneiki	36.00 B	39.70 B	44.32 D	44.54 C	6.67 C	7.05 C
Coratina	46.67 A	60.00 A	48.65 B	51.45 B	10.42 A	13.72 A

5% probability level ($p < 0.05$).

Table 13.
Olive yield, and oil content and yield of the tested cultivars.

Genotypes and parental cv.	Crossing combination	Receptors	Leaf shape	Fruit shape index	Fruit weight	Flesh/seed	Oil content
13	Chemlali × Toffahi	Chemlali	Elliptic-lanceolate	Ovoid	M	M	H
14	Chemlali × Toffahi	Chemlali	Elliptic-lanceolate	Ovoid	M	M	H
15	Aggizi × Arbequin	Aggizi	Elliptic-lanceolate	Ovoid	VH	M	M
53	Manzanillo × Aggizi	Manzanillo	Elliptic-lanceolate	Spherical	H	M	H
68	Toffahi × Kalamata	Toffahi	Elliptic-lanceolate	Elongated	H	M	M
69	Toffahi × Kalamata	Toffahi	Elliptic-lanceolate	Ovoid	H	H	M
75	Arbequin × Aggizi	Arbequin	Elliptic-lanceolate	Ovoid	M	L	H
77	Arbequin × Aggizi	Arbequin	Lanceolate	Spherical	M	L	M
79	Arbequin × Aggizi	Arbequin	Elliptic-lanceolate	Spherical	H	M	H
81	Arbequin × Aggizi	Arbequin	Elliptic-lanceolate	Spherical	M	L	H
85	Arbequin × Aggizi	Arbequin	Elliptic-lanceolate	Ovoid	L	L	H
86	Arbequin × Aggizi	Arbequin	Elliptic-lanceolate	Spherical	M	L	M
Chemlali			Elliptic-lanceolate	Elongated	L	L	H
Aggizi			Elliptic-lanceolate	Spherical	VH	M	VL
Toffahi			Elliptic-lanceolate	Spherical	VH	M	VL
Arbequin			Elliptic-lanceolate	Elongated	L	L	H
Kalamata			Elliptic-lanceolate	Elongated	VH	M	H
Manzanillo			Elliptic-lanceolate	Ovoid	H	L	H

H: high; M: medium; L: low; VH: very high; VL: very low.

Table 14.
Some morphological traits and oil content of the evaluated genotypes and parental cultivars.

DOI: http://dx.doi.org/10.5772/intechopen.1008350

Variety	66	69	Arbosana	Oleana	Koroneikii	Arbequina	LSD (0.05)
Parameter:							
Fruit weight (g)	3.16 ± 0.55	3.11 ± 0.33	3.71 ± 0.25	2.18 ± 0.20	3.21 ± 0.40	2.77 ± 0.22	0.44
Seed weight (g)	0.88 ± 0.08	0.74 ± 0.10	0.81 ± 0.10	0.71 ± 0.10	0.72 ± 0.13	0.47 ± 0.09	0.12
Flesh weight (g)	2.26 ± 0.74	2.36 ± 0.25	2.93 ± 0.28	1.43 ± 0.50	2.36 ± 0.20	1.57 ± 0.25	0.35
Fruitmoisture (%)	62.05 ± 2.21	62.17 ± 2.33	63.48 ± 2.74	64.07 ± 2.44	58.85 ± 2.22	64.53 ± 1.85	12.1
Fruit oil (%) D.W.	33.75 ± 1.65	35.59 ± 1.33	40.18 ± 2.00	32.90 ± 1.09	42.74 ± 1.05	32.93 ± 1.25	6.05
Pigments (mg/kg):							
Chlorophyll	6.5 ± 0.35	7.2 ± 0.4	6.2 ± 0.4	5.3 ± 0.3	7.2 ± 0.3	5.4 ± 0.3	1.05
Carotenoids	6.66 ± 0.2	6.4 ± 0.4	6.45 ± 0.3	6.25 ± 0.5	6.68 ± 0.4	6.39 ± 0.4	1.07
α-Tocopherol	55.42 ± 0.8	63.61 ± 0.8	57.95 ± 0.90	64.11 ± 0.80	72.63 ± 0.90	56.22 ± 0.90	10.26

Table 15.
Chemical composition (on dry basis) and pigments chlorophyll and carotene (mg/kg) of the six investigated olive varieties.

Author details

Ahmed Mohamed Taha* and Hamdy El-Houssainy Khalifa
Soils, Water and Environment Research Institute (SWERI), Egypt Agricultural Research Center (ARC), Egypt

*Address all correspondence to: ahmedtaha362@yahoo.com

References

[1] Kühlert M, Klingen J, Gröne K, Hennes L, Terrapon-Pfaff J, Jamea EM. Pathways towards a Green Economy in Egypt - Analyzing Decarbonization and Resource Efficiency Pathways along Value Chains to Support Green Private Sector Development in Egypt. Germany: Wuppertal Institute; 2024

[2] World Bank. Egypt: Country Climate and Development Report. World Bank; 2022. Available from: https://www.worldbank.org/en/country/egypt/publication/egypt-country-andclimate-development-report

[3] IMF. African Countries with the Highest Gross Domestic Product (GDP) in 2022 (in Billion U.S. Dollars). Statista; 2023a. Available from: https://www.statista.com/statistics/1120999/gdp-of-african-countries-by-country/

[4] Cooper AF, Antkiewicz A, Shaw TM. Lessons from BRICSAM about south-north relations at the start of the 21st century: Economic size trumps all Else. International Studies Review. 2007;**9**(4):675-687

[5] Ouda S, Zohry AE-H. Crop rotation: An approach to save irrigation water under water scarcity in Egypt. In: WatSave Innovative Water Management Award 2015. 2015

[6] El-Gindy AM. Sustainable use of agricultural resources program. In: Program Board Meeting, USA. 2011, in Monitoring and Evaluation of Irrigation Management Projects in Egypt. JARQ. 2015;**49**(2):111-118. Availble from: http://www.jircas.affrc.go.jp

[7] Seada T, Mohamed R, Fletscher T, Abouleish H, Abouleish-Boes M. The Future of Agriculture in Egypt: Comparative Study of Organic and Conventional Food Production Systems in Egypt. Egypt: Heliopolis University for Sustainable Development; 2016

[8] Food and Agricultural Organization (FAO). AQUASTAT Country Profile–Egypt. Food and Agriculture Organization of the United Nations (FAO); 2016. Available from: https://www.fao.org/3/i9729en/I9729EN.pdf

[9] Food and Agricultural Organization (FAO). Fertilizer Use by Crop in Egypt. Rome, Italy: FAO, Land & Water Development Division; 2005

[10] Medany MA. Climate Change: Impacts and Responses for Sustainable Agriculture in Egypt. CIHEAM, Watch Letter no. 37; Sep 2016

[11] Allam MN, Allam GI. Water resources in Egypt: Future challenges and opportunities. Water International. 2007;**32**(2):205-218

[12] Djuma H, Bruggeman A, Eliades M, Lange MA. Non-conventional water resources research in semi-arid countries of the Middle East. Desalination and Water Treatment. 2016;**57**(5):2290-2303

[13] Khadr M, Zeidan BA, Elnmer A. On-farm water management in the Nile Delta. In: The Nile Delta. 2017. pp. 325-344. DOI: 10.1007/698_2016_56

[14] Abd Ellah RG. Water resources in Egypt and their challenges, Lake Nasser case study. Egyptian Journal of Aquatic Research. 2020;**46**(1):1-12

[15] Negm AM, Sakr S. An overview of groundwater resources in Nile Delta aquifer. In: Negm AM, editor. Groundwater in the Nile Delta. Cham: Springer International Publishing; 2019. pp. 3-44

[16] Abutaleb KAA, Mohammed AHE-S, Ahmed MHM. Climate change impacts, vulnerabilities and adaption measures for Egypt's Nile Delta. Earth Systems and Environment. 2018;**2**(2):183-192

[17] Egypt Environmental Affairs Agency (EEAA). Egypt Second National Communication under the United Nations Framework Convention on Climate Change. EEAA; 2010

[18] World Bank. Accelerating the Transformation of Egypt's Agri-Food System: Impact of Investments in Selected Value-Chains. World Bank; 2021. Available from: worldbank.org/curated/en/259611625118209081/Egypt-Arab-Republic-of-Programmatic-Support-to-MALR

[19] Waha1 K, Krummenauer L, Adams S, Aich V, Baarsch F, Coumou D, et al. Climate change impacts in the Middle East and Northern Africa (MENA) region and their implications for vulnerable population groups. Regional Environmental Change. 2017;**17**(6):1623-1638. DOI: 10.1007/s10113-017-1144-2

[20] Namdar R, Karami E, Keshavarz M. Climate change and vulnerability: The case of MENA countries. ISPRS International Journal of Geo-Information. 2021;**10**(11):794. DOI: 10.3390/ijgi10110794

[21] Nassar SZ, Ahmed YN, Siam GM, Nayera YS, Shaimaa HS. Analysis of climate change effects on food security in Egypt using IMPACT model. Egyptian Journal of Agricultural Economy. 2021;**31**(3):1002-1011

[22] Fraga H, Moriondo M, Leolini L, Santos JA. Mediterranean olive orchards under climate change: A review of future impacts and adaptation strategies. Agronomy. 2021;**2021**(11):56. DOI: 10.3390/agronomy11010056

[23] Newton C, Terral J-F, Ivorra S. The Egyptian olive (Olea europaea subsp. Europaea) in the later first millennium BC: Origins and history using the morphometric analysis of olive stones. Antiquity. 2006;**80**:405-414

[24] IOC (International Olive Council). Policies – Egypt. 2012. 12 p. Available from: http://www.internationaloliveoil.org/estaticos/view/130-survey-and-assessment-division

[25] Abou-Zaid FO. Olive oil and rural development in Egyptian deserts. In: Management and Development of Agricultural and Natural Resources in Egypt's Desert. Springer Nature; 2021. pp. 451-490. DOI: 10.1007/978-3-030-73161-8_17

[26] Mahmoud DHI, El-Dnasury MF, El-Noby AK. An analytical study for production and marketing of the olive crop in Egypt (Case study in Matrouh governorate). SVU-International Journal of Agricultural Sciences. 2023;**5**(1):57-74

[27] Teimaa EA. The economic analysis of the most important factors affecting on the olive sector in Egypt. Journal of Agricultural Economics & Social Sciences. 2023;**14**(2):57-62. DOI: 10.21608/jaess.2023.188025.1143

[28] Nassar A. Comparison of micro irrigation systems for olive trees. Misr Journal of Agricultural Engineering. 2009;**26**(1):149-169

[29] Attalla AM, Abdel-Sattar M, Mahrous AE, Abdel-Azeez AA. Olive trees productivity in response to supplemental irrigation under North-Western coastal conditions in Egypt. American-Eurasian Journal of Agricultural & Environmental Sciences. 2011;**11**(5):609-615

[30] El-Taweel AA, Farag AA. Effect of different water quantities on growth and fruit quality of olive trees under new

reclaimed lands conditions. Journal of Plant Production, Mansoura University. 2016;**7**(1):13-27

[31] Sary D, Elsokkary I. Effect of irrigation water regime in presence of organic or biological fertilizer on olive trees. Egyptian Journal of Soil Science. 2019;**59**(1):67-84

[32] Ali HAM, Gowda AM, Farrag HM, Radwan EMA. Performance of some olive (*Olea europaea* L) cultivars grown under saline stress conditions in newly reclaimed soils. Assiut Journal of Agriculture Science. 2022;**53**(4):39-54

[33] Alowaiesh BF, El-Mansy AB, El-Alakmy DA, Rehema EA, El-Moneim DABD. Biochemical, physiological, and molecular responses of diverse olive cultivars to different irrigation regimes. Notulae Botanicae Horti Agrobotanici Cluj-Napoca. 2023;**51**(4):13395

[34] Taha AM, Khalifa HE. Productivity of olive oil (Coratina variety) in response to irrigation treatments in sandy soil. Irrigation and Drainage. 2023;**2023**:1-17

[35] Khalifa HE, Taha AM. Response of table olive (*Aggizi cultivar*) to different amounts of irrigation in sandy soil. Journal of Soils and Crops. 2023;**33**(2):240-252

[36] El-Fouly MM, El-Taweel AA, Osman IMS, El-Din IS, Shaban SHA. Nutrient removal from different parts of Koroneiki olive trees grown in sandy soils as a base of fertilizer recommendation in Egypt. British Journal of Applied Science & Technology. 2014;**4**(11):1718-1728

[37] Abd-Alhamid N, Laila FH, Hassan HSA, Abdelhafez AA. Effect of mineral and bio-fertilization on yield and fruit quality of Manzanillo olive trees. International Journal of ChemTech Research CODEN (USA). 2015;**8**(11):63-73

[38] Abd El-Migeed MMM, El-Naggar MAA, Afifi AA, El-Shawadfy MA. Response of olive trees (cv. Koroneiki) to algae extract sprays and its impact on growth and productivity under saline conditions. Middle East Journal of Agriculture. 2018;**7**(1):34-40

[39] Abd El-Razek E, Lila FH, Eman SE-H, Shahin MFM. Effect of soil application of humic acid and bio-humic on yield and fruit quality of "Kalamata" olive trees. Bulletin of the National Research Centre. 2020, 2020;**44**:73

[40] Ikram SE-D, Mikhail EG, Osman IMS. Evaluation of some olive hybrids derived from a breeding program. Journal of Agricultural Science, Mansoura University. 2009;**34**(7):8111-8128

[41] Hegazi AA. Performance of 12 introduced olive cultivars under Egyptian conditions. Research Journal of Agriculture and Biological Sciences. 2012;**8**(2):98-107

[42] Fayek MA, Abdel-Mohsen MA, Sanaa IL, El-Sayed SM. Morphological, agronomical and genetic characterization of Egyptian olive clones compared with the international cultivars. Egyptian Journal of Horticulture. 2014;**41**(1):59-82

[43] Omran MA. Morphological and productivity of some olive genotypes derived from a breeding program and comparing with their parental cultivars American-Eurasian. Journal of Agriculture and Environmental Sciences. 2021;**21**(1):22-38

[44] Arafat SM, Abd El-Baset WS, Ellabban AA. The quality of olive oil extracted from some olive varieties cultivated by highly intensive in Egypt. Egyptian Journal of Chemistry. 2022;**65**(8):407-417

Chapter 5

Olive Oil and Diseases

Maria Trapali

Abstract

The Mediterranean diet, which has been followed in Greece for centuries, is a modern object of study and analysis. Its main ingredients are olive oil and table olives. It is now characterized as a diet with an extremely positive effect on human health. The consumption of extra virgin olive oil and table olives, basic components of the Mediterranean diet, offers significant benefits to human health. Olive oil is a source of energy and monounsaturated, "good" fats, and a rich source of antioxidants, especially extra virgin olive oil, which helps in the absorption of fat-soluble vitamins (A, D, and E) from food, helps reduce the risk of cardiovascular diseases, and contributes, in the context of the Mediterranean diet, to the better management of various other diseases such as type 2 diabetes mellitus, cancer, neurodegenerative diseases, and Alzheimer's disease. It provides antioxidants and many of the fats necessary for good development during neonatal and childhood.

Keywords: olive oil, Mediterranean diet, antioxidants, T2DM, Alzheimer's disease

1. Introduction

The Mediterranean diet, which has been followed in Greece for centuries, is a modern object of study and analysis. Its main ingredients are olive oil and table olives. It is now characterized as a diet with an extremely positive effect on human health. The consumption of extra virgin olive oil and table olives, basic components of the Mediterranean diet, offers significant benefits to human health [1, 2].

In general, the consumption of fats and oils is essential for the normal functioning of the body since they provide energy, vitamins, and other nutrients, while at the same time they add flavor to our food. However, both the type and the amount we consume are of particular importance for our health. The term virgin olive oil refers to those olive oils that have been obtained from olives using only mechanical methods and without any chemical or biochemical treatment. In other words, olive oil is produced exclusively by processing the olives in the olive mill and without any admixture with oils of different origins. This category includes extra virgin, virgin, and low-quality olive oils. The difference between the three subcategories of virgin olive oil is determined by a series of chemical measurements and by tasting analysis. Referring only to the gustatory properties, that is, the olfactory and gustatory impression of olive oil, a virgin olive oil without any gustatory defects and with a fruity sensation is classified as extra virgin and belongs to the higher category of gustatory and nutritional quality. Nuclear oil—is obtained from the kernel and pulp of the olive tree after special processing. Often, virgin olive oil can be added to it [3].

Acidity is one of the most important parameters for classifying olive oil and evaluating the quality of extra virgin olive oil. It measures the concentration in % of fatty acids in the free state and is therefore not involved in the formation of triglycerides, the chemical compounds of which olive oil is mainly composed. In order for an olive oil to be classified as extra virgin, the legislation stipulates that its free acidity must not exceed 0.8%. Olive oil with an acidity of between 0.8% and 2% is classified as virgin, while oil with an acidity of 2% or more is not consumed unless refined. Other evaluation characteristics include peroxides, ultraviolet absorption indexes, antioxidants, waxes, and organoleptic characteristics such as fruitiness [3–5].

Extra virgin olive oil should have low acidity, high antioxidants, low peroxides, low wax, and UV absorption indices, and, above all, be fruity. It is also rich in monounsaturated fatty acids, which are among the "good" fats to be consumed daily. It does not come from seeds but is the juice from the fruit of the olive tree, which retains the abundance of beneficial substances as it is not subjected to any further processing. Olive oil is the key ingredient of the famous Mediterranean diet, while extra virgin olive oil (EVOO) is the only cooking oil that is 100% natural and does not undergo any chemical processing. There are studies that show that eating olive oil can help prevent strokes, heart disease, Alzheimer's, diabetes, osteoporosis, and even cancer [6]. Consumption of EVOO can reduce levels of systemic inflammation. Specifically, the molecule oleocanthal, by acting as an inhibitor of specific enzymes, is involved in preventing disease and maintaining good body health. Studies suggest that extra virgin olive oil can reduce pain, inflammation, and oxidative stress associated with rheumatoid arthritis. It is also very beneficial for cardiovascular health. It contains polyphenols, which reduce the risk of heart disease, low density lipoprotein (LDL) cholesterol levels, total cholesterol levels, and contribute positively to the prevention of atherosclerotic plaque formation in the arteries [2].

Frequent consumption of extra virgin olive oil reduces the daily dose of medication in hypertensive patients. The plant phenols in olive oil are beneficial in the fight against obesity by reducing food intake while regulating proper cell function, a process that helps reduce the inflammation associated with obesity. Also, the polyphenols found in extra virgin olive oil improve fat metabolism, tissue sensitivity to insulin, and the regulation of blood glucose levels.

EVOO may also help improve liver function, which is associated with metabolic syndrome and protects against non-alcoholic fatty liver disease, oxidative stress, and the inflammatory response [2]. Studies show that olive oil can destroy bacteria associated with ulcers and stomach cancer. Strains of the bacterium *Helicobacter pylori* have shown some resistance to antibiotics. In tests conducted with extra virgin olive oil, it was shown to be able to eliminate all strains of the bacteria, including those with antibiotic resistance. Polyphenols also appear to reduce the accumulation of Alzheimer-related amyloid proteins, leading to reduced chances of cognitive deficit.

In conclusion, extra virgin olive oil helps lowering bad cholesterol (LDL); maintain good cholesterol (HDL); preventing heart disease and certain types of cancer; antiaging and wrinkle prevention, since olive oil contains vitamin E; against Alzheimer's disease, according to a new American study; in the healing of gastric ulcers and rheumatoid arthritis and is therapeutic in the diet against diabetes.

2. Composition of the olive oil

The type and quantity of the components contained in olive oil vary depending on the variety, the altitude, the harvest period and the extraction method used. It contains mainly oleic acid (up to 83%), with smaller amounts of other fatty acids, including linoleic acid (up to 21%) and palmitic acid (up to 20%) [7–9].

2.1 Oleic acid

Oleic acid is a fatty acid found naturally in various animal and vegetable fats and oils. It is an odorless, colorless oil, although commercial samples may be yellowish. It is chemically classified as a monounsaturated omega-9 fatty acid and briefly given the lipid number 18:1 cis-9. It has the chemical formula $CH_3(CH_2)_7CH{=}CH(CH_2)_7COOH$ (**Figure 1**). Its name is derived from the Latin word oleum, meaning oil.

2.2 Linoleic acid

Linoleic acid (LA) is a polyunsaturated omega-6 fatty acid, with an 18-carbon chain and two double bonds in a cis configuration. It is briefly referred to as "18:2 (n-6)" or "18:2 cis-9,12" (**Figure 2**). Linoleic acid is one of the two essential fatty acids that the human body cannot synthesize from other food components.

Linoleic acid is used in the biosynthesis of arachidonic acid, which produces prostaglandins, leukotrienes (LTA, LTB, and LTC), and thromboxane (**Figure 3**).

2.3 Palmitic acid

Palmitic acid or hexanoic acid, is a common saturated fatty acid. Its chemical formula is $CH_3(CH_2)_{14}COOH$, and C16:0 for short (**Figure 4**).

2.4 Vitamin E

Alpha-tocopherol (vitamin E) is the main tocopherol in olive oil and is of most interest from a dietary point of view because it has the highest biological value (**Figure 5**).

2.5 Polyphenols

A health claim for olive oil is made if it contains at least 5 mg of hydroxytyrosol and its derivatives (oleuropein and tyrosol complex) per 20 g of olive oil. In order for this

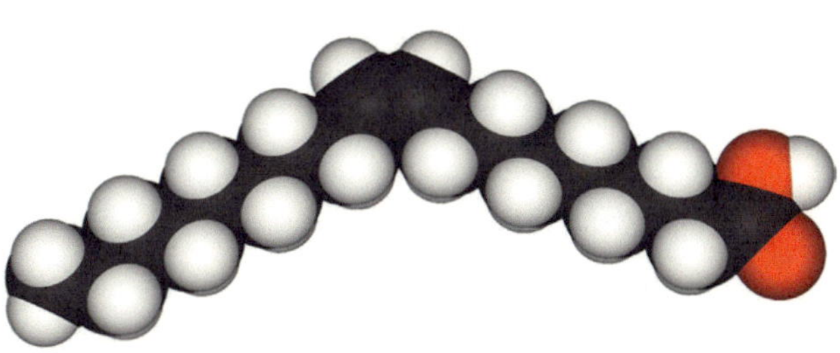

Figure 1.
Oleic acid molecule. https://commons.wikimedia.org/w/index.php?search=oleic+acid&title=Special:MediaSearch&go=Go&type=image

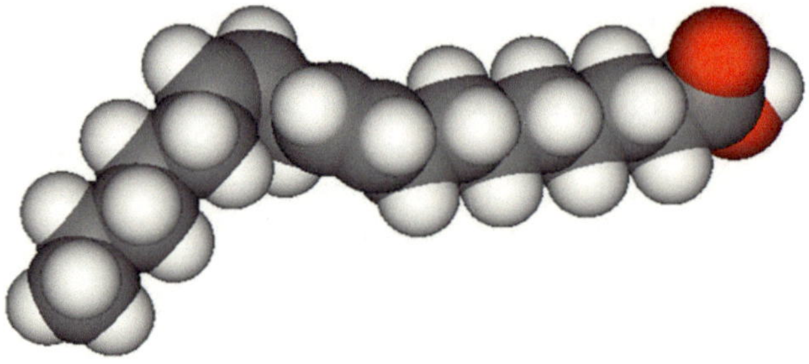

Figure 2.
Linoleic acid molecule. https://commons.wikimedia.org/w/index.php?search=%CE%9B%CE%99%CE%9D%CE%95%CE%9B%CE%91%CE%AA%CE%9A%CE%8C+%CE%9F%CE%9E%CE%8E&title=Special:MediaSearch&go=Go&type=image

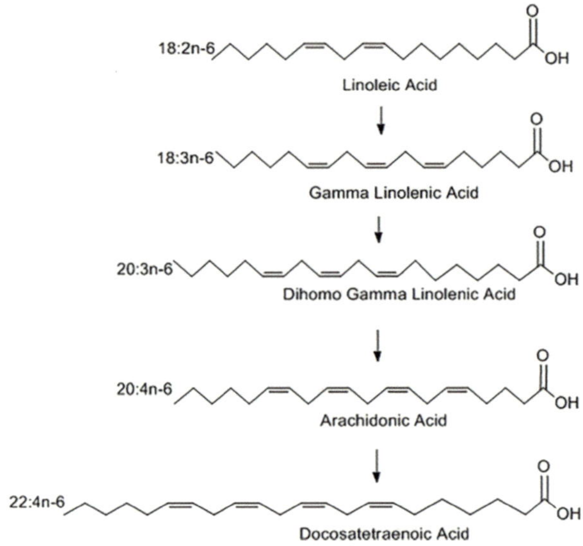

Figure 3.
Linoleic acid metabolism. https://commons.wikimedia.org/w/index.php?search=%CE%9B%CE%99%CE%9D%CE%95%CE%9B%CE%91%CE%AA%CE%9A%CE%8C+%CE%9F%CE%9E%CE%8E&title=Special:MediaSearch&go=Go&type=image

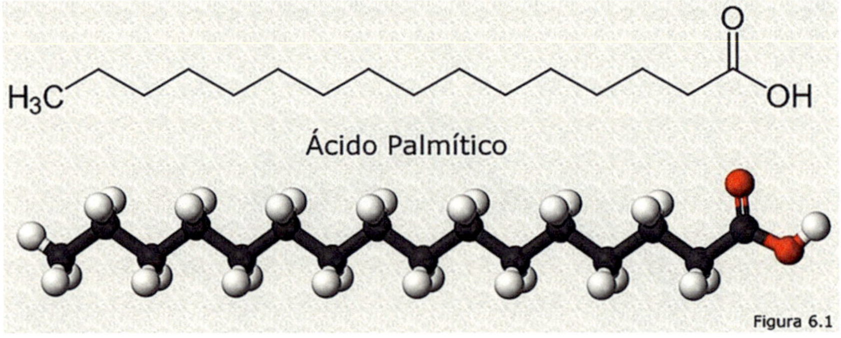

Figure 4.
Palmitic acid molecule. https://commons.wikimedia.org/w/index.php?search=PALMITIC+ACID&title=Special:MediaSearch&go=Go&type=image

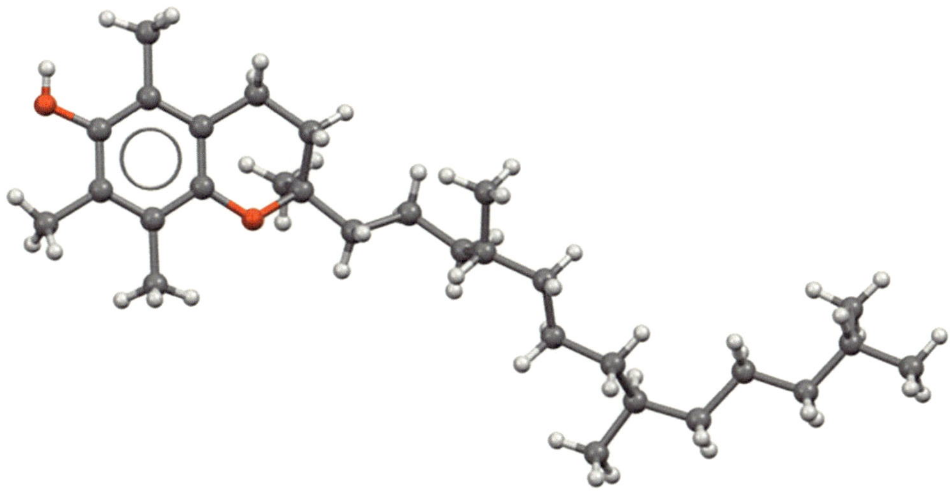

Figure 5.
Vitamin E molecule. https://commons.wikimedia.org/w/index.php?search=%CE%91+%CE%A4%CE%9F%CE%9A%CE%9F%CE%A6%CE%95%CE%A1%CE%8C%CE%9B%CE%97&title=Special:MediaSearch&go=Go&type=image

claim to be used, the consumer should be informed that the beneficial effects are ensured by a daily intake of 20 g of olive oil (reference portion of olive oil). Tyrosol, hydroxytyrosol (**Figure 6**), and oleuropein (**Figure 7**) are three typical phenolic compounds present in olives in free or conjugated forms as aglycones. Other phenolic constituents include oleacin and oleocanthal, which are derivatives of hydroxytyrosol and tyrosol, respectively, ligstroside, pinoresinol, acetoxypinorencinol, and verbascoside [10, 11].

Tyrosol is a highly stable compound that prevents the oxidation of "bad" LDL cholesterol, a fact of great importance since this oxidation is one of the main steps in the process of atherosclerosis [9].

Hydroxytyrosol is produced by the hydrolysis of oleuropein that occurs during olive ripening. Oleuropein accumulates in the leaves and fruits of the olive as a defense mechanism against pathogens and herbivores [12].

Hydroxytyrosol (HT) represents one of the main polyphenol contents of EVOO and has anti-inflammatory and anti-teratogenic activity, improving the lipid profile, reducing oxidative stress, and activating inflammatory cells. It also acts on the expression of peroxisome proliferator-activated receptors (PPAR) γ and α, which reduces adipocyte size [13]. In addition, oleuropein, another antioxidant found in unripe olive fruits and olive leaves, has also been associated with improved inflammatory parameters in various models of inflammation, in addition to antiproliferative and anticancer properties that induce the apoptosis process of cancer cells, mainly in the colon region [14, 15].

Oleocanthal (**Figure 8**) is a phenylethanoid, a natural phenolic compound found in extra virgin olive oil. It appears to be responsible for the burning sensation that

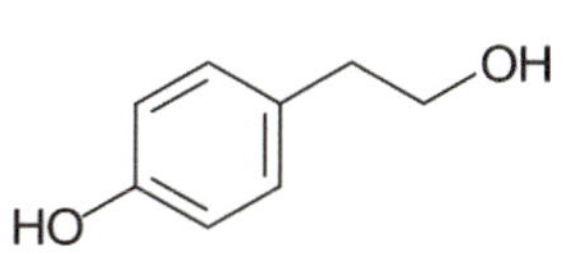

Figure 6.
Tyrosol molecule. https://commons.wikimedia.org/w/index.php?search=%CE%A4%CE%A5%CE%A1%CE%9F%CE%A3%CE%8C%CE%9B%CE%97&title=Special:MediaSearch&go=Go&type=image

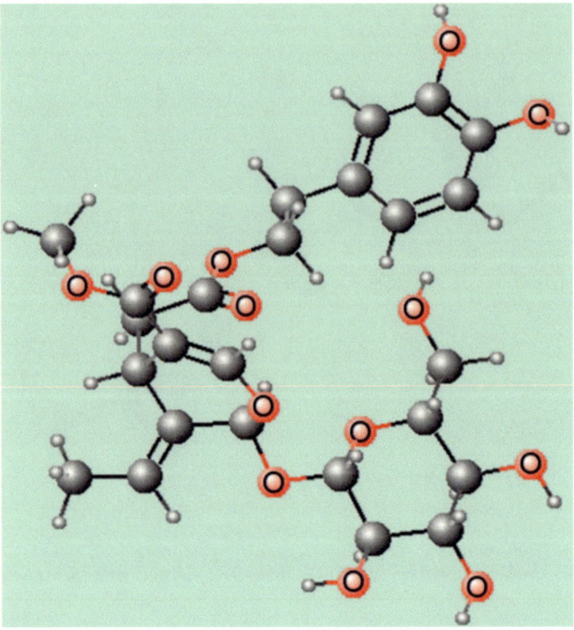

Figure 7.
Oleuropein molecule. https://commons.wikimedia.org/w/index.php?search=Oleuropein&title=Special:MediaSearch&go=Go&type=image

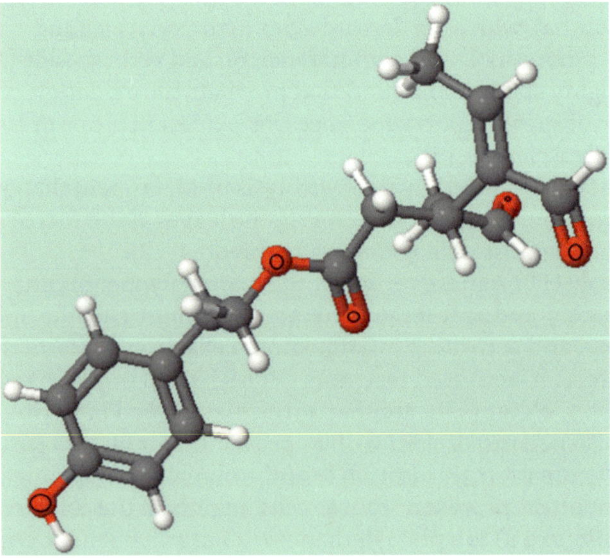

Figure 8.
Oleocanthal molecule. https://commons.wikimedia.org/w/index.php?search=%CE%B5%CE%BB%CE%B1%CE%B9%CE%BF%CE%BA%CE%B1%CE%BD%CE%B8%CE%AC%CE%BB%CE%B7&title=Special:MediaSearch&go=Go&type=image

occurs at the back of the throat when such oil is consumed. Oleocanthal is an ester of tyrosol and its chemical structure is related to that of oleuropein, which is also found in olive oil [16].

2.6 Flavonoids

Luteolin (**Figure 9**) and apigenin (**Figure 10**) are the most important flavonoids in olive oil.

2.7 Chlorophyll - carotenoids

The color of olive oil is due to two categories of pigments—chlorophylls, including their derivatives, and carotenoids (lutein and beta-carotene). Chlorophylls and their derivatives are responsible for the greenish brown color of the oils, while carotenoids are responsible for the yellow color.

Carotenoids play an important role in the stability, freshness, and nutritional value of extra virgin olive oil (EVOO). However, the carotenoid content in EVOO changes over time as a function of olive ripening and deterioration events. A reliable indicator of quality is the ratio between the two most abundant carotenoids, namely lutein and beta-carotene, the latter being more rapidly degraded. Research studies have focused on developing a method using Raman spectroscopy to quantify the ratio of lutein to beta-carotene

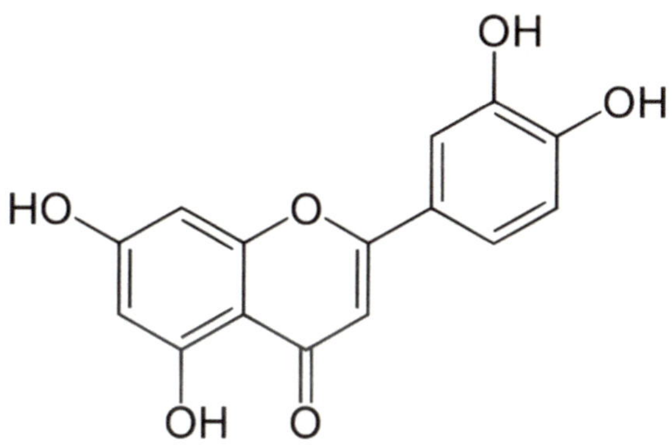

Figure 9.
Luteolin molecule. https://commons.wikimedia.org/w/index.php?search=%CE%BB%CE%BF%CF%85%CF%84%CE%B5%CE%BF%CE%BB%CE%AF%CE%BD%CE%B7&title=Special:MediaSearch&go=Go&type=image

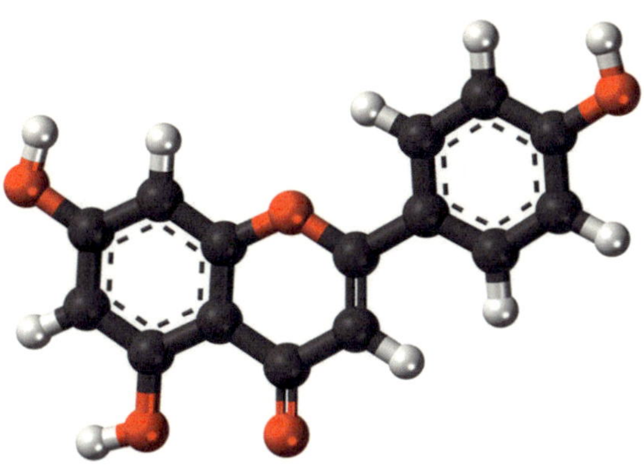

Figure 10.
Apigenin molecule. https://commons.wikimedia.org/w/index.php?search=%CE%B1%CF%80%CE%B9%CE%B3%CE%B5%CE%BD%CE%AF%CE%BD%CE%B7&title=Special:MediaSearch&go=Go&type=image

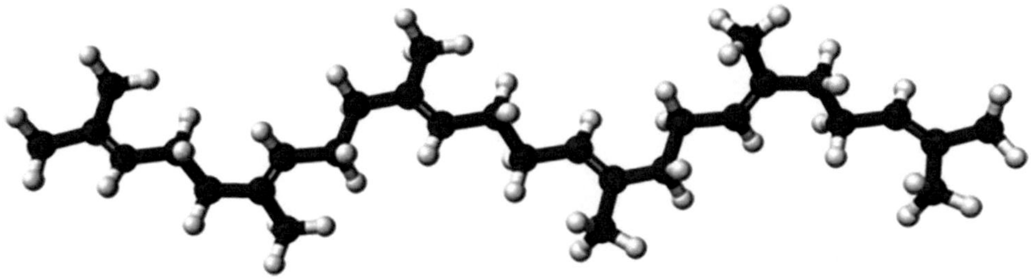

Figure 11.
Squalene molecule. https://commons.wikimedia.org/w/index.php?search=%CF%83%CE%BA%CE%BF%CF%85%CE%B1%CE%BB%CE%AD%CE%BD%CE%B9%CE%BF&title=Special:MediaSearch&go=Go&type=image

in extra virgin olive oil. This ratio is an important indicator of the quality and authenticity of olive oil [11, 17]. The study by Benito de Valle-Prieto et al. investigates the enrichment of virgin olive oil with lutein and zeaxanthin extracted from *Spinacia oleracea* (spinach). The research aimed to enhance the nutritional value of virgin olive oil by incorporating these carotenoids, which are known for their antioxidant properties and potential eye health benefits. The process involved extracting lutein and zeaxanthin from spinach and then incorporating them into olive oil. The enriched olive oil was then analyzed to determine the stability of the added compounds and potential health benefits. The study concluded that virgin olive oil could be effectively fortified with lutein and zeaxanthin, maintaining the quality of the oil and providing additional health benefits [18].

2.8 Squalene

Squalene (**Figure 11**) is the main hydrocarbon in virgin olive oil and is an unsaturated terpene compound that covers more than 50% of the unsaponifiable part of olive oil. Olive oil contains up to 300 times more squalene than other vegetable oils and up to 5000 times more than some plant foods. Also, olive oil, because it is not subject to heat or other processing, contains much higher amounts of squalene than processed seed oils or other oils [6, 19].

3. Olive oil and diseases

3.1 Olive oil and cardiovascular disease

Cardiovascular disorders (**Figure 12**) are listed as one of the leading causes of death worldwide and, when associated with risk factors such as dyslipidemia, arterial hypertension and inflammation, increase the development of the disease and associated metabolic symptoms. Olive oil consumption led to a reduction in cardiovascular risk parameters such as inflammatory cytokine (interleukin-6), vascular cell adhesion molecule (VCAM), and intercellular adhesion molecule 1 (ICAM-1), and an increase in high-density lipoprotein levels (HDL-c) and reduced LDL-c levels. Endothelial dysfunction results in the accumulation of adhesion molecules that cause vascular damage. Adhesion molecules are synthesized in endothelial cells and their production is stimulated by inflammatory cytokines such as interleukin 1 (IL-1) and tumor necrosis factor (TNF-α). In healthy individuals, suppression of such adhesion molecules occurs and the reduction in sick individuals suggests fewer cardiovascular events [13, 20].

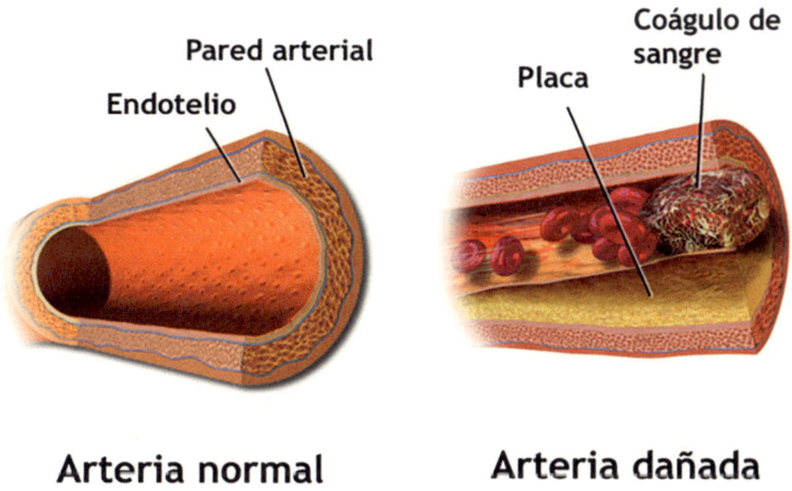

Figure 12.
Development of atherosclerotic plaque in an artery. https://commons.wikimedia.org/w/index.php?search=%CE%91%CE%98%CE%97%CE%A1%CE%A9%CE%9C%CE%86%CE%A4%CE%A9%CE%A3%CE%97&title=Special:MediaSearch&go=Go&type=image

In adults with metabolic syndrome, consumption of 10 mL of extra virgin olive oil (EVOO) per day reduced abdominal obesity and LDL cholesterol [21].

The phenolic compounds present in EVOO act as antioxidants by preventing and reducing cardiovascular disease events through reducing pro-inflammatory cytokines, improving endothelial function, inhibiting lipid peroxidation caused by free radicals or metals, and inhibiting LDL-c activity as well as HDL-c oxidation, as shown in research studies. Indeed, the administration of 25 mL (phenol content: 366 mg/kg caffeic acid equivalent) of olive oil to adults resulted in a reduction in systolic blood pressure and LDL [22]. In patients with hypercholesterolemia, administration of EVOO +500 mg/kg polyphenols EVOO and thyme (1:1) caused a decrease in oxidized LDL [23]. Similarly, the administration of 25 mL of olive oil to adult overweight women reduced body fat and diastolic pressure [24].

At the same time, it seems that the consumption of olive oil positively affects the secretion of adiponectin, a hormone formed in fat cells that has antidiabetic and anti-inflammatory action, that is, it is a cardioprotective hormone. In obese people, its levels are reduced, which causes insulin resistance, arterial hypertension, and the development of inflammation and atherosclerosis, which are predisposing factors for the development of cardiovascular disease [16, 25, 26]. **Figure 13** shows the Relative risk and 95% CI for fatal and nonfatal cardiovascular disease associated with a 25 g per day increase in olive oil consumption.

3.2 Olive oil and gut biota health

Consumption of EVOO has been associated with promoting gut health by promoting the development of a higher biodiversity of gut bacteria and influencing the ratio of Firmicutes/Bacteroidetes. There has also been an increase in the genus Clostridium XIVa, whose cellular components and their metabolites, such as butyrate, secondary bile acids, and indole propionic acid, play a probiotic role mainly through activation of intestinal epithelial cells, strengthening the intestinal barrier, interaction with the immune system, and development of anti-inflammatory activity [26–28].

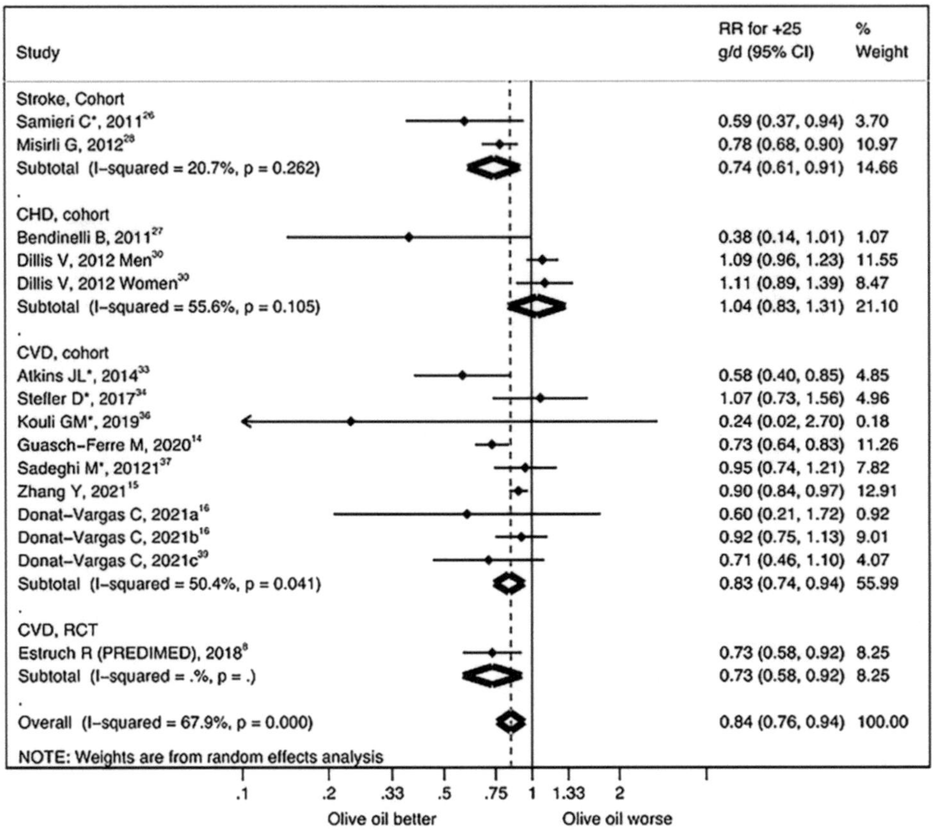

Figure 13.
Relative risk and 95% CI for fatal and nonfatal cardiovascular disease associated with a 25 g per day increase in olive oil consumption. https://commons.wikimedia.org/w/index.php?search=OLIVE+OIL+AND+DISEASES&title=Special:MediaSearch&go=Go&type=image

Serum levels of lipopolysaccharide (LPS), zonulin (gut permeability marker), glucose, insulin, and glucagon-like peptide 1 (GLP1) were measured in 20 patients with impaired fasting glucose (IFG) and 20 healthy subjects (HS) according to gender and age. Patients with IFG showed higher levels of LPS and zonulin; in the group of healthy volunteers, meal intake was accompanied by significant increases in blood glucose, insulin, and GLP1 with no changes in blood LPS and zonulin. Two hours after taking a meal containing EVOO, IFG patients showed a lower increase in blood glucose, an increase in blood insulin and GLP1, and a significant decrease in LPS and zonulin compared with IFG patients who did not take EVOO, and LPS was found to be positively correlated with blood glucose and zonulin and negatively correlated with blood insulin [10, 29, 30].

3.3 Olive oil and inflammation

Inflammation is mainly divided into acute and chronic. Acute inflammation is the first reaction of the immune system, while chronic inflammation is a consequence of inflammation that persists and leads to the deregulation of various metabolic pathways with sequelae of chronic diseases. In the inflammatory response, leukocytes and mast cells are present in areas of damage and directed to a "respiratory burst" as a result of enhanced oxygen uptake and therefore enhance the production and circulation of reactive oxygen species (ROS) in the damaged area. Locally at the site of injury, leukocytes enter the injury site from the circulation, which is activated by various chemokines and cytokines secreted by dendritic cells and macrophages. At this stage, the leukocytes release further cytokines and inflammatory mediators. Neutrophils contain granules rich in lysozyme, matrix metalloproteinases, and myeloperoxidase (MPO) which are released into the foreign or autoantigen, leading to its destruction with phagocytosis, the release of ROS and cytokines such as interleukin-1 (IL-1), interleukin-6 (IL-6), and tumor necrosis factor (TNF-α). The next line of defense is the lymphocytes, including T- and B-lymphocytes, which have an important role in mediating inflammation through many complex mechanisms such as lymphocyte co-stimulation, cytokine secretion, and the production of antibodies and immune complexes [12, 13]. There is a balance between ROS/free radical formation and endogenous antioxidant defense mechanisms in the normal and healthy state of the body. When this is disrupted, it can lead to oxidative stress and associated damage that can result in injury to all vital cellular components, such as DNA, proteins, and lipids of membrane proteins and cell death (**Figure 14**). As a result, it can cause many diseases, including diabetes, cardiovascular disease, inflammation, cancer, degenerative diseases, ischemia, and anemia [31].

The oleocanthal contained in olive oil has strong anti-inflammatory properties. It acts similarly to non-steroidal anti-inflammatory drugs by reducing the production of inflammatory molecules such as cytokines. Frequent consumption of olive oil can help reduce chronic inflammation.

Administration of EVOO enriched with 500 mg/kg polyphenols to adults with hypercholesterolemia caused an increase in C-reactive protein (CRP) and immunoglobulin IgA. CRP secreted mainly by the liver (also by adipose tissue, macrophages,

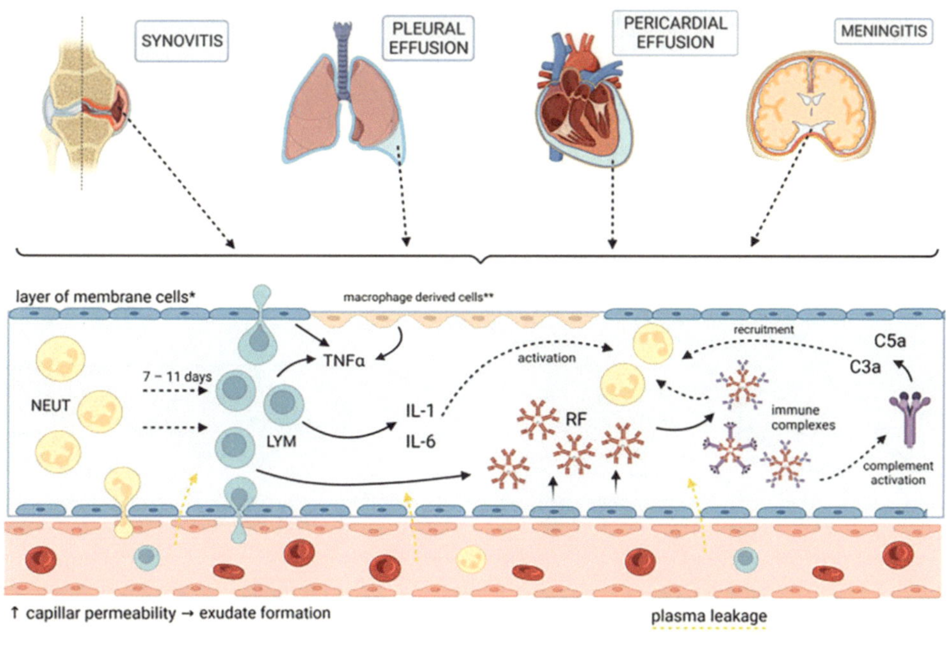

Figure 14.
The inflammation processes. https://commons.wikimedia.org/w/index.php?search=%CF%86%CE%BB%CE%B5%CE%B3%CE%BC%CE%BF%CE%BD%CE%AE&title=Special:MediaSearch&go=Go&type=image

and endothelial cells) is a marker of acute inflammation and a marker of chronic inflammation used in cardiology as a predictor of infarction. CRP joins the damaged cells and bacteria to initiate activation of the complement system, secretion of cytokines, recruitment of monocytes and other leukocytes from the blood, and apoptosis of damaged cells [32]. In adults with impaired fasting glucose levels, administration of 10 g EVOO caused a decrease in oxidized LDL and NADPH oxidase 2 (Nox2) levels [31].

However, the administration of olive oil (487 ppm phenols and 389 ppm triterpenes) to healthy adults reduced the levels of tumor necrosis factor (TNF-α) and interleukin-6 (IL-6) [33]. In experimental animals, EVOO administration decreased the levels of vascular cell adhesion molecule-1 (VCAM-1), interferon gamma (IFN-γ), and IL-6 (**Figure 15**) [26].

3.4 Olive oil and diabetes mellitus (T2DM)

Diabetes mellitus (**Figure 16**) is a metabolic disorder characterized by an increase in blood sugar concentration (hyperglycemia) and impaired glucose metabolism, either as a result of reduced insulin secretion or due to reduced sensitivity of cells to insulin [24, 34, 35]. When we refer to the existence of insulin resistance, we mean the reduced sensitivity of peripheral tissues, and especially the muscles to the action of insulin in terms of glycose hemostasis. Insulin resistance can be due to autoantibodies to the insulin receptor, such as in type B insulin resistance syndrome, or mutations in the insulin receptor insulin receptor. In most cases, however, the mechanisms are complex and involve multiple pathways in the intracellular transmission of the insulin message as well as the interaction of mediators, metabolic or hormonal as in polycystic fibrosis syndrome ovarian polycystic fibrosis, metabolic syndrome, and obesity [34]. Olive oil can improve insulin sensitivity through the effect of the monounsaturated fatty acids it contains, which help improve cellular function and increase insulin response. This mechanism is particularly important for people with metabolic syndrome or T2DM [36].

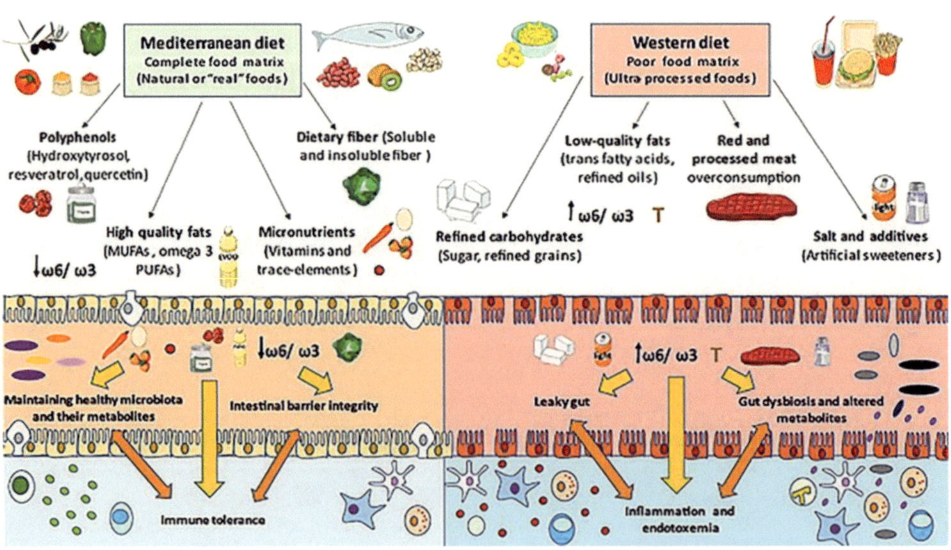

Figure 15.
A general overview of the main nutritional components modulating both gut microbiota and immune system. https://commons.wikimedia.org/w/index.php?search=Mediterranean+diet&title=Special:MediaSearch&go=%CE%9C%CE%B5%CF%84%CE%AC%CE%B2%CE%B1%CF%83%CE%B7&uselang=el&type=image

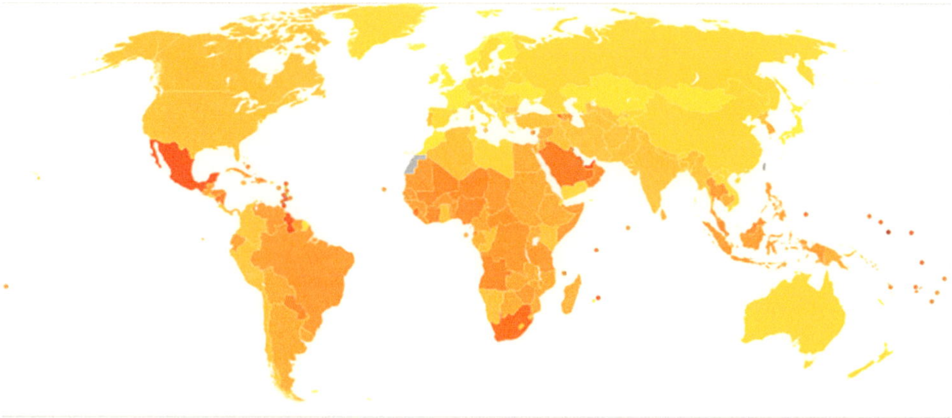

Figure 16.
Diabetes mellitus world map—DALY—WHO2004. https://commons.wikimedia.org/w/index.php?search=diabetes+mellitus&title=Special:MediaSearch&go=%CE%9C%CE%B5%CF%84%CE%AC%CE%B2%CE%B1%CF%83%CE%B7&uselang=el&type=image

Chronic low-grade inflammation and oxidative stress are also implicated in the pathogenesis of T2DM [37–39]. In a study involving 3230 participants with T2DM with a follow-up of 3.2 years, EVOO consumption delayed the introduction of new antidiabetic treatments in patients [40]. A meta-analysis was performed to find the optimal macronutrient composition in healthy diets. Diets high in cis-mono unsaturated fatty acids (MUFA) can improve metabolic risk factors among patients with T2DM, compared to diets high in carbohydrate (CHO) or polyunsaturated fatty acids (PUFA) [36].

Also, a study in overweight patients with T2DM showed that consumption of a polyphenol-rich extra virgin olive oil (HP-EVOO) significantly reduced fasting plasma glucose and glycosylated hemoglobin (HbA1c) levels, as well as body weight and body mass index. There was also a significant reduction in transaminases and visfatin, a hormone that has been considered as a possible link between visceral fat deposition and the development process of diabetes mellitus [41, 42].

Amylin is an insulin-like substance and when it accumulates in pancreatic cells it creates amyloid, which has a direct link to the pathogenesis of T2DM. In particular, it appears that oleuropein aglycone inhibits amylin accumulation and its toxic effect on pancreatic cells. Also, antidiabetic functions have been attributed to hydroxytyrosol and oleuropein. However, the detailed molecular mechanisms of how it affects beta-cell functions remain poorly understood [43–45].

The INS-1 cell line is derived from a transplantable insulinoma created by X-ray in rats. Because INS-1 cells contain a high concentration of insulin and respond to changes in glucose levels, they are often used to study beta cell function. INS-1E cells were exposed to 10 μM EVOO for 24 h and survival, insulin biosynthesis, glucose-stimulated insulin secretion (GSIS) and activation of intracellular signaling (protein kinase B [AKT] and cAMP response element binding protein [CREB]) were evaluated. Hydroxytyrosol, tyrosol, and apigenin increased β-cell proliferation and insulin biosynthesis, and apigenin and lutein enhanced GSIS. Apigenin was the most effective compound and was also able to activate beneficial intracellular signaling. In conclusion, this study shows that hydroxytyrosol, tyrosol, and apigenin enhance beta-cell health, suggesting that EVOO or supplements enriched with these compounds may improve insulin secretion and promote glycemic control in patients with T2DM [7, 46].

In *in vitro* experiments to study the antidiabetic effect, it was found that hydroxytyrosol had the strongest inhibitory activity of α-glucosidase with IC50 values of 150 μM, while the type of inhibition was non-competitive. The same type of inhibition was also found in oleuropein. These results suggest that hydroxytyrosol and oleuropein are two potential effective inhibitors of α-glucosidase for the management of postprandial hyperglycemia [47, 48].

Thirty patients with IFG were randomly allocated to a meal containing or not containing 10 g EVOO in a crossover design. Before 60 and 120 min after lunch, blood samples were taken to measure glucose, insulin, peptide-1 (GLP1), dipeptidyl peptidase-4 (DPP4) activity, triglycerides (TG), total cholesterol, HDL-cholesterol, and Apo B-48. The meal containing EVOO was associated with a decrease in glucose ($p = 0.009$) and DPP4 activity ($p < 0.001$) and a significant increase in insulin ($p < 0.001$) and GLP-1 ($p < 0.001$) compared to the meal without EVOO. In addition, the meal containing EVOO showed a significant decrease in triglycerides ($p = 0.002$) and Apo B-48 ($p = 0.002$) compared to the meal without EVOO. Total cholesterol and HDL-cholesterol levels did not change significantly between the two groups [49].

One of the complications of diabetes is diabetes kidney disease. Lutein levels were significantly lower in these patients, negatively correlated with body mass index, glycosylated hemoglobin, fasting blood glucose, triglycerides, and positively correlated with high-density lipoprotein cholesterol in addition, long-term administration of lutein to the retina of Ins2Akita/+ mice caused the abolition of retinal inflammation and retinal vessel armoring [50].

3.5 Olive oil and cognitive functions

Clinical trials and experimental data in review studies suggest that olive oil phenolics may play an important role in slowing or preventing cognitive decline in aging populations, highlighting their potential as a dietary intervention to maintain cognitive health and combat neurodegenerative disorders such as Alzheimer's (**Figure 17**) [51–53].

Another study examines the relationship between adoption of the Mediterranean diet and cognitive function in adults. The results of the study show that the Mediterranean diet is associated with positive effects on cognitive functions, providing protective benefits for brain health, particularly in relation to aging, and reducing the risk of neurodegenerative diseases [54]. Another study investigating the effects of EVOO on blood-brain barrier (BBB) function in people with mild cognitive impairment showed that EVOO consumption significantly improved BBB function, possibly providing a protective mechanism against cognitive decline and the progression of mild cognitive impairment (MCI) to more severe forms of dementia [55].

In mouse brain endothelial cell culture, oleocanthal treatment increased the expression and activity of P-glycoprotein (P-gp) and LDL lipoprotein receptor-related protein-1 (LRP1). Administration of oleocanthal extract of extra virgin olive oil to wild-type C57BL/6 mice enhanced clearance of β-amyloid (1-40) peptide labeled with iodine 125, (125) I-Aβ40, from the brain and increased the brain efflux index (BEI %) method; brain efflux index characterizes an efflux transport system for substrates from the brain to circulating blood along the blood-brain barrier). The findings experimentally support that the risk of Alzheimer's disease is reduced with EVOO consumption through enhancing the clearance of Ab from the brain [56].

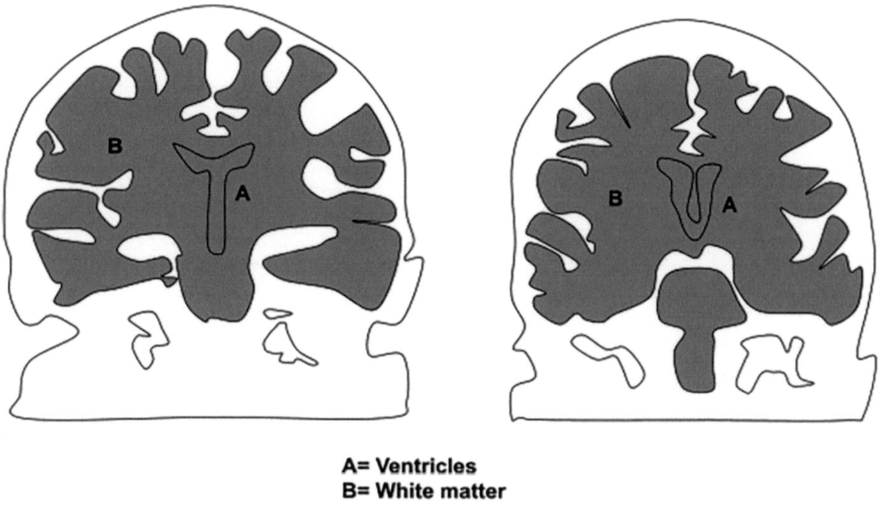

Figure 17.
The connection between olive oil and Alzheimer's disease, focusing on the protective potential of olive oil for cognitive health. https://commons.wikimedia.org/w/index.php?search=Mediterranean+diet&title=Special:MediaSearch&go=%CE%9C%CE%B5%CF%84%CE%AC%CE%B2%CE%B1%CF%83%CE%B7&uselang=el&type=image

3.6 Olive oil and cancer

A systematic review and meta-analysis showed that the protective effect of olive oil was more pronounced for certain types of cancer (**Figure 18**), such as breast cancer and colorectal cancer. At the cellular level, oleocanthal inhibits the growth and tendency of breast or prostate cancer cells to metastasize. Even in small amounts, oleocanthal and thus extra virgin olive oil have a potential therapeutic role in breast and prostate cancer. Also, a review of various preclinical studies and experimental models highlights the efficacy of olive oil polyphenols in reducing cancer cell viability and enhancing the effects of conventional cancer therapies. The possible mechanisms of action, are through induction of apoptosis, interference with cancer cell growth and division, inhibition of metastasis, and regulation of cell cycle and survival signaling [57, 58].

A research study combining the administration of olive oil polyphenols (hydroxytyrosol, oleuropein, and tyrosol) with traditional chemotherapy or targeted therapies could improve therapeutic outcomes and possibly reduce side effects [59].

4. Conclusions

The consumption of extra virgin olive oil and table olives, basic components of the Mediterranean diet, offers significant benefits to human health. Their positive effect, based on modern research studies, is attributed mainly to polyphenols, and more specifically to phenols such as hydroxytyrosol and tyrosol. More generally, polyphenolic antioxidants have the ability to scavenge free radicals and reduce certain chelating reactions: reactive oxygen-containing ions (free radicals) must be removed from cells continuously to maintain proper metabolism. From research to date, polyphenolic antioxidants can reduce inflammatory effects (such as coronary heart disease), delay the aging process, and protect proteins and DNA. Olive products (mainly table olives

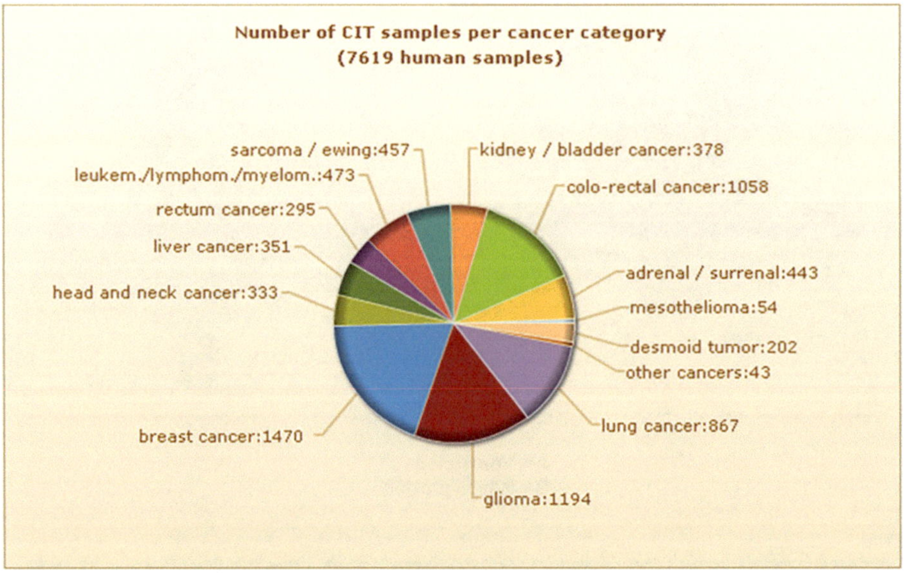

Figure 18.
Cancer types. https://commons.wikimedia.org/w/index.php?search=types+of+cancer&title=Special:MediaSearch&go=%CE%9C%CE%B5%CF%84%CE%AC%CE%B2%CE%B1%CF%83%CE%B7&uselang=el&type=image

and olive oil) are classified as medical foods, due to their nutritional benefits and protective properties against cardiovascular diseases, age-related diseases, neurodegenerative disorders, and other diseases.

European Regulation 432/2012 mentions the health claim "Olive oil polyphenols contribute to the protection of blood lipids from oxidative stress". This health claim concerns olive oils that contain at least 5 mg of hydroxytyrosol and its derivatives (e.g., oleoeuropein and tyrosol complex) per 20 g of olive oil (REGULATION EU No. 432/2012).

Conflict of interest

The author declares no conflict of interest.

Author details

Maria Trapali
Laboratory of Chemistry, Biochemistry and Cosmetic Science, Department of Biomedical Medicine, University of West Attica, Attiki, Greece

*Address all correspondence to: ymaria@uniwa.gr

References

[1] Guasch-Ferré M, Liu G, Li Y, Sampson L, Manson JE, Salas-Salvadó J, et al. Olive oil intake and risk of cardiovascular disease and mortality among U.S. adults. Journal of the American College of Cardiology. 2020;**75**(15):1729-1739. DOI: 10.1016/j.jacc.2020.02.036

[2] Trichopoulou A, Lagiou P, Kuper H. The Mediterranean diet: A history of health. Public Health Nutrition. 2000;**3**(1):5-8. DOI: 10.1017/S1368980000000107

[3] Mueller T. Extra Virginity: The Sublime and Scandalous World of Olive Oil. W.W. Norton & Company; 2012. ISBN: 978-0-393-34361-8

[4] Boskou D, Talbot G, editor. Olive Oil and Health. In: Specialty Oils and Fats in Food and Nutrition. 1st edition. Chapter: 1. Woodhead Publishing; 2015. pp. 3-33

[5] Lycopolus E. The Olive Oil and Vinegar Lover's Cookbook. Touch Wood Editions; 2015. ISBN: 9781771513029

[6] Kiritsakis A, Keceli MT, Kiritsakis K. Olive oil. In: Shahidi F, editor. Bailey's Industrial Oil and Fat Products. Wiley, Co; 2020. pp. 1-38. DOI: 10.1002/047167849X.bio029.pub2

[7] Boskou D, Blekas G, Tsimidou M. Olive oil composition. In: Owusu-Apenten EGV, editor. Olive Oil. Food Lipids: Chemistry, Biotechnology, and Analysis. CRC Press; 2006. pp. 65-92. DOI: 10.1201/9781439832028.pt2

[8] Tsimidou M, Boskou D. The role of phenolic compounds in the oxidation of olive oil. Journal of the American Oil Chemists' Society. 2003;**80**(8):649-654. DOI: 10.1007/s11746-003-0765-2

[9] Covas MI, Miró-Casas E, Fitó M, Farré-Albaladejo M, Gimeno E, Marrugat J, et al. Bioavailability of tyrosol, an antioxidant phenolic compound present in wine and olive oil, in humans. Drugs Under Experimental and Clinical Research. 2003;**29**(5-6):203-206

[10] Bellumori M, Cecchi L, Innocenti M, Clodoveo ML, Corbo F, Mulinacci N. The EFSA health claim on olive oil polyphenols: Acid hydrolysis validation and total hydroxytyrosol and tyrosol determination in Italian virgin olive oils. Molecules. 2019;**24**(11):2179. DOI: 10.3390/molecules24112179

[11] Anselmi C, Portarena S, Baldacchini C, Proietti S, Leonardi L, Brugnoli E. One drop only: Easy and rapid Raman evaluation of β-carotene in olive oil and its relevance as an index of olive fly attack. Food Chemistry. 2022;**393**:133340. DOI: 10.1016/j.foodchem.2022.133340

[12] Charoenprasert S, Mitchell A. Factors influencing phenolic compounds in table olives (*Olea europaea*). Journal of Agricultural and Food Chemistry. 2012;**60**(29):7081-7095. DOI: 10.1021/jf3017699

[13] Hohmann CD, Cramer H, Michalsen A, Kessler C, Steckhan N, Choi K, et al. Effects of high phenolic olive oil on cardiovascular risk factors: A systematic review and meta-analysis. Phytomedicine. 2016;**22**:631-640. DOI: 10.1016/j.phymed.2015.03.019

[14] Giner E, Recio MC, Ríos JL, Cerdá-Nicolás JM, Giner RM. Chemopreventive

effect of oleuropein in colitis-associated colorectal cancer in c57bl/6 mice. Molecular Nutrition & Food Research. 2016;**60**:242-255. DOI: 10.1002/mnfr.201500605

[15] Foscolou A, Critselis E, Panagiotakos D. Olive oil consumption and human health: A narrative review. Maturitas. 2018;**118**:60-66. DOI: 10.1016/j.maturitas.2018.10.013

[16] Scoditti E, Massaro M, Carluccio MA, Pellegrino M, Wabitsch M, Calabriso N, et al. Additive regulation of adiponectin expression by the mediterranean diet olive oil components oleic acid and hydroxytyrosol in human adipocytes. PLoS One. 2015;**10**:e0128218. DOI: 10.1371/journal.pone.0128218

[17] Portarena S, Anselmi C, Leonardi L, Proietti S, Bizzarri AR, Brugnoli E, et al. Lutein/β-carotene ratio in extra virgin olive oil: An easy and rapid quantification method by Raman spectroscopy. Food Chemistry. 2023;**404**(Pt B):134748. DOI: 10.1016/j.foodchem.2022.134748

[18] Benito de Valle-Prieto M, Delgado-Adámez J, Gil MV, Martillanes S, Franco MN, Martín-Vertedor D. Virgin olive oil enriched with lutein-zeaxanthin from *Spinacia oleracea*. Journal of Oleo Science. 2017;**66**(5):463-468. DOI: 10.5650/jos.ess16189

[19] Gaforio JJ, Sánchez-Quesada C, López-Biedma A, del Carmen Ramírez-Tortose M, Warleta F. Molecular aspects of squalene and implications for olive oil and the Mediterranean diet. In: The Mediterranean Diet. Cambridge, MA, USA: Academic Press; 2015. pp. 281-290

[20] Gungor ZB. Vascular inflammation and genetic predisposition as risk factors for cardiovascular diseases. In: Nagpanl ML, editor. Cholesterol—Good, Bad, and the Heart. 1st ed. London, UK, London, UK: IntechOpen; 2018

[21] Venturini D, Simão ANC, Urbano MR, Dichi I. Effects of extra virgin olive oil and fish oil on lipid profile and oxidative stress in patients with metabolic syndrome. Nutrition. 2015;**31**:834-840. DOI: 10.1016/j.nut.2014.12.016

[22] Martín-Peláez S, Castañer O, Konstantinidou V, Subirana I, Muñoz-Aguayo D, Blanchart G, et al. Effect of olive oil phenolic compounds on the expression of blood pressure-related genes in healthy individuals. European Journal of Nutrition. 2017;**56**:663-670. DOI: 10.1007/s00394-015-1110-z

[23] Martín-Peláez S, Mosele JI, Pizarro N, Farràs M, de la Torre R, Subirana I, et al. Effect of virgin olive oil and thyme phenolic compounds on blood lipid profile: Implications of human gut microbiota. European Journal of Nutrition. 2017;**56**:119-131. DOI: 10.1007/s00394-015-1063-2

[24] Cândido FG, Valente FX, Silva LE, Coelho OGL, Peluzio MCG, Alfenas RCG. Consumption of extra virgin olive oil improves body composition and blood pressure in women with excess body fat: A randomized, double-blinded, placebo-controlled clinical trial. European Journal of Nutrition. 2017;**57**:7

[25] Castro-Barquero S, Lamuela-Raventós RM, Doménech M, Estruch R. Relationship between Mediterranean dietary polyphenol intake and obesity. Nutrients. 2018;**10**:1523. DOI: 10.3390/nu10101523

[26] Katsarou AI, Kaliora AC, Chiou A, Kalogeropoulos N, Papalois A, Agrogiannis G, et al. Amelioration of oxidative and inflammatory status

in hearts of cholesterol-fed rats supplemented with oils or oil-products with extra virgin olive oil components. European Journal of Nutrition. 2016;**55**:1283-1296. DOI: 10.1007/s00394-015-0947-5

[27] Hidalgo M, Prieto I, Abriouel H, Villarejo AB, Ramírez-Sánchez M, Cobo A, et al. Changes in gut microbiota linked to a reduction in systolic blood pressure in spontaneously hypertensive rats fed an extra virgen olive oil-enriched diet. Plant Foods for Human Nutrition. 2018;**73**:1 6. DOI: 10.1007/s11130-017-0650-1

[28] Deiana M, Serra G, Corona G. Modulation of intestinal epithelium homeostasis by extra virgin olive oil phenolic compounds. Food & Function. 2018;**9**:4085-4099. DOI: 10.1039/C8FO00354H

[29] Serum levels of lipopolysaccharides (LPS), zonulin, a marker of gut permeability, glucose, insulin and glucagon-like peptide 1 (GLP1) were measured in 20 patients with impaired fasting glucose (IFG) and 20 healthy subjects (HS) matched for sex and age

[30] Bartimoccia S, Cammisotto V, Nocella C, Del Ben M, D'Amico A, Castellani V, et al. Extra virgin olive oil reduces gut permeability and metabolic endotoxemia in diabetic patients. Nutrients. 2022;**14**(10):2153. DOI: 10.3390/nu14102153

[31] Carnevale R, Pastori D, Nocella C, Cammisotto V, Bartimoccia S, Novo M, et al. Gut-derived lipopolysaccharides increase post-prandial oxidative stress via Nox2 activation in patients with impaired fasting glucose tolerance: Effect of extra-virgin olive oil. European Journal of Nutrition. 2019;**58**:2. DOI: 10.1007/s00394-018-1718-x

[32] Martín-Peláez S, Castañer O, Solà R, Motilva MJ, Castell M, Pérez-Cano FJ, et al. Influence of phenol-enriched olive oils on human intestinal immune function. Nutrients. 2016;**8**:213. DOI: 10.3390/nu8040213

[33] Sanchez-Rodriguez E, Biel-Glesson S, Fernandez-Navarro JR, Calleja MA, Espejo-Calvo JA, Gil-Extremera B, et al. Effects of virgin olive oils differing their bioactive compound contents on biomarkers of oxidative stress and inflammation in healthy adults: A randomized double-blind controlled trial. Nutrients. 2019;**11**:561. DOI: 10.3390/nu11030561

[34] Bourkoula A, Konsta E, Papadopoulou A, Trapali M. Lipidic classes involved in diabetes mellitus: A review. Novel Research in Sciences. 2021;**8**(2):000685. DOI: 10.31031/NRS.2021.08.000685

[35] Trapali M. Berberine and type II diabetes: A mini review. Endocrinology, Diabetes and Metabolism Journal. 2024;**8**(2):1-7

[36] Qian F, Korat AA, Malik V, Hu FB. Metabolic effects of monounsaturated fatty acid-enriched diets compared with carbohydrate or polyunsaturated fatty acid-enriched diets in patients with type 2 diabetes: A systematic review and meta-analysis of randomized controlled trials. Diabetes Care. 2016;**39**:1448-1457. DOI: 10.2337/dc16-0513

[37] Trapali M. Oxidic degradation of lipids in patients with type II diabetes. Review Clinical Pharmacology and Pharmacokinetics. 2021;**35**(2):75-77

[38] Trapali M. Antioxidant activity in patients with type II diabetes. Review Clinical Pharmacology and Pharmacokinetics. 2022;**36**(1):6

[39] Lourenço A, Freitas C, Timóteo M, Soares M, Figueiredo J, Osório N, et al. Laboratory assessment of the effects of AGA@4life multidisciplinary intervention on the inflammatory profile. MMPs, and TIMPs in a Geriatric Population Healthcare. 2024;**12**(5):509. DOI: 10.3390/healthcare12050509

[40] Basterra-Gortari FJ, Ruiz-Canela M, Martínez-González MA, Babio N, Sorlí JV, Fito M, et al. Effects of a Mediterranean eating plan on the need for glucose-lowering medications in participants with type 2 diabetes: A subgroup analysis of the PREDIMED trial. Diabetes Care. 2019;**42**:1390-1397. DOI: 10.2337/dc18-2475

[41] Schwingshackl L, Strasser B. High-MUFA diets reduce fasting glucose in patients with type 2 diabetes. Annals of Nutrition & Metabolism. 2012;**60**:33-34. DOI: 10.1159/000335162

[42] Santangelo C, Filesi C, Varì R, Scazzocchio B, Filardi T, Fogliano V, et al. Consumption of extra-virgin olive oil rich in phenolic compounds improves metabolic control in patients with type 2 diabetes mellitus: A possible involvement of reduced levels of circulating visfatin. Journal of Endocrinological Investigation. 2016;**39**:1295-1301. DOI: 10.1007/s40618-016-0506-9

[43] Wu L, Velander P, Liu D, Xu B. Olive component Oleuropein promotes β-cell insulin secretion and protects β-cells from amylin amyloid-induced cytotoxicity. Biochemistry. 2017;**56**:5035-5039. DOI: 10.1021/acs.biochem.7b00199

[44] Chaari A. Inhibition of human islet amyloid polypeptide aggregation and cellular toxicity by oleuropein and derivatives from olive oil. International Journal of Biological Macromolecules. 2020;**162**:284-300. DOI: 10.1016/j.ijbiomac.2020.06.170

[45] Lee H, Im SW, Jung CH, Jang YJ, Ha TY, Ahn J. Tyrosol, an olive oil polyphenol, inhibits ER stress-induced apoptosis in pancreatic β-cell through JNK signaling. Biochemical and Biophysical Research Communications. 2016;**469**:748-752. DOI: 10.1016/j.bbrc.2015.12.036

[46] Marrano N, Spagnuolo R, Biondi G, Cignarelli A, Perrini S, Vincenti L, et al. Effects of extra virgin olive oil polyphenols on beta-cell function and survival. Plants. 2021;**10**:286. DOI: 10.3390/plants10020286

[47] Hadrich F, Bouallagui Z, Junkyu H, Isoda H, Sayadi S. The α-glucosidase and α-amylase enzyme inhibitory of hydroxytyrosol and oleuropein. Journal of Oleo Science. 2015;**64**:835-843. DOI: 10.5650/jos. ess15026

[48] Bozzetto L, Alderisio A, Giorgini M, Barone F, Giacco A, Riccardi G, et al. Extra-virgin olive oil reduces glycemic response to a high-glycemic index meal in patients with type 1 diabetes: A randomized controlled trial. Diabetes Care. 2016;**39**:518-524. DOI: 10.2337/dc15-2189

[49] Carnevale R, Loffredo L, Del Ben M, Angelico F, Nocella C, Petruccioli A, et al. Extra virgin olive oil improves post-prandial glycemic and lipid profile in patients with impaired fasting glucose. Clinical Nutrition. 2017;**36**:782-787. DOI: 10.1016/j.clnu.2016.05.016

[50] Trapali M. Lutein in chronic diseases: A mini review. Review of Clinical Pharmacology And Pharmacokinetics, International. 2024;**38**(1):47-56. DOI: 10.61873/YBCS9028

[51] Boronat A, Serreli G, Rodríguez-Morató J, Deiana M, de la Torre R. Olive oil phenolic compounds' activity against age-associated cognitive decline: Clinical

and experimental evidence. Antioxidants (Basel). 2023;**12**(7):1472. DOI: 10.3390/antiox12071472

[52] Alkhalifa AE, Al-Ghraiybah NF, Kaddoumi A. Extra-virgin olive oil in Alzheimer's disease: A comprehensive review of cellular, animal, and clinical studies. International Journal of Molecular Sciences. 2024, 2024;**25**(3):1914. DOI: 10.3390/ijms25031914

[53] Fazlollahi A, Motlagh Asghari K, Aslan C, Noori M, Nejadghaderi SA, Araj-Khodaei M, et al. The effects of olive oil consumption on cognitive performance: A systematic review. Frontiers in Nutrition. 2023;**10**:1218538. DOI: 10.3389/fnut.2023.1218538

[54] Hardman RJ, Kennedy G, Macpherson H, Scholey AB, Pipingas A. Adherence to a Mediterranean-style diet and effects on cognition in adults: A qualitative evaluation and systematic review of longitudinal and prospective trials. Frontiers in Nutrition. 2016;**3**:22. DOI: 10.3389/fnut.2016.00022

[55] Kaddoumi A, Denney TS Jr, Deshpande G, Robinson JL, Beyers RJ, Redden DT, et al. Extra-virgin olive oil enhances the blood-brain barrier function in mild cognitive impairment: A randomized controlled trial. Nutrients. 2022;**14**(23):5102. DOI: 10.3390/nu14235102

[56] Abuznait A, Qosa H, Busnena B, Sayed K, Amal K. Olive-oil-derived oleocanthal enhances β-amyloid clearance as a potential neuroprotective mechanism against Alzheimer's disease: *In vitro* and *in vivo* studies. ACS Chemical Neuroscience. 2013;**4**(6):973-982

[57] Pei T, Zhang X, Chen H, Zhang L, Zhao L, Wang J, et al. Anti-cancer effects of olive oil polyphenols and their underlying mechanisms. Oncotarget. 2016;**7**(28):43475-43491. DOI: 10.18632/oncotarget.8694

[58] Markellos C, Ourailidou M-E, Gavriatopoulou M, Halvatsiotis P, Sergentanis TN, Psaltopoulou T. Olive oil intake and cancer risk: A systematic review and meta-analysis. PLoS One. 2022;**17**(1):e0261649. DOI: 10.1371/journal.pone.0261649

[59] Torić J, Marković AK, Brala CJ, Barbarić M. Anticancer effects of olive oil polyphenols and their combinations with anticancer drugs. Acta Pharmaceutica. 2019;**69**(4):461-482. DOI: 10.2478/acph-2019-0052

Chapter 6

Olive Oil Tourism: Innovative or Traditional Form of Rural Tourism?

Maria Lúcia Pato

Abstract

Olive oil tourism is a unique and rapidly growing segment of agritourism, particularly popular in Mediterranean countries such as Portugal, where olive oil production has deep cultural and historical roots. At the heart of this form of tourism is the opportunity for visitors to explore the olive-growing process, engage in tastings, and immerse themselves in the traditions surrounding one of the world's most ancient food products. However, modern innovation is playing a pivotal role in revitalizing these traditions, making olive oil tourism not just a journey into the past but a dynamic exploration of the future. Based on a qualitative analysis of five farms of olive oil tourism, findings indicate the focus on tradition and values of the past as the main ingredient to promote the offer of olive oil tourism. Through this blend of old and new, olive oil tourism not only preserves heritage but also paves the way for a sustainable and vibrant future. Innovation through tradition is indeed a path that must be considered by scholars and practitioners in the promotion of olive oil tourism and related products.

Keywords: olive oil tourism, tradition, endogenous resources, innovation, rurality

1. Introduction

Olive oil tourism is observed as a specific and new type of rural tourism linked to agriculture, the tradition of the rural space, and the production of olive oil [1]. Its expansion started in many regions as an alternative to complement agricultural income, promoting familiar and small-scale agriculture and local farmers [2], and help to promote olive oil and the respective territory characterized by olive trees [3]. On the one hand, tourists seek authenticity, tradition, history, the nostalgia of the past, and to emerge in the rural culture of the territory. But on the other hand, tourists also seek novelty and creative reinventions of somewhat forgotten [4]. It is a mixture of the past with the present, in order to promote an authentic and innovative experience for tourists.

While innovation is important to ensure a competitive advantage in any touristic product [5], innovation through tradition (ITT) is the main ingredient to promote, in a sustainable way, the development of olive oil territory. This may allow for the discovery of values, practices, and forgotten abilities connected to specific traditions that may sustain unique and/or distinct products or services of the rural space [5].

Accordingly, new products and services may arise from the process of recombining endogenous and cultural tradition as well as the tradition of some long-standing small family firms and the adoption of cutting-edge technologies [6].

Despite the recognition of the importance of innovation in rural tourism, to the best of our knowledge, to date, research that focuses ITT is scarce, particularly in a new product as olive oil tourism. Based on this gap, the purpose of this paper is to gain insights concerning the relationship between tradition and innovation in olive oil tourism and observe some examples in Portugal, one of the main producers of olive oil in the world. Data used are based on a qualitative analysis of five websites farms that offer experiences in olive oil tourism. Findings indicate the focus on tradition and values of the past as the main ingredient to promote the offer of olive oil tourism.

The paper consisted of five parts after the introduction. Section 2 provides a review of the "connections and requirements of olive oil tourism" as well as the concept of ITT. The region under analysis and the methodological approach are explained in Section 3, while Section 4 presents and discusses the results of the websites analysis. Finally, in the last section, Section 5, the study's main conclusions are summarized, the study's limitations are pointed out, and possible areas for future research are put forward.

2. Literature review

2.1 Connections and requirements of olive oil tourism

Olive oil tourism is a form of rural tourism that centers around the production and cultural significance of olive oil. This type of tourism typically involves diverse connections with public or private spaces. Often tourists can visit olive groves and olive oil mills or learn about the process of growing olives, extracting olive oil, and its various uses (gastronomic, religious, and/or medicinal). For this reason, it intersects other types of tourism such as gastronomic tourism [7], cultural tourism, health and well-being tourism [3], religious tourism [8, 9], or even industrial tourism [10]. In this sense, it can have many potentialities for the development of these regions, both at an environmental, economic, and sociocultural level [1]. Some key experiences of oil tourism involved the following:

- Guided tours to olive groves: visitors can take guided tours of olive groves to see how olives are cultivated [10, 11]. These tours often include explanations of different olive varieties, cultivation methods, and harvesting techniques.

- Visits to olive oil mills: tours of olive oil mills provide insights into the extraction process, including traditional and modern methods of pressing olives to produce oil [8, 10].

- Shopping: visitors often have the opportunity to purchase olive oil and related products directly from the producers, ensuring they get authentic and high-quality items [12].

- Olive oil tastings: tastings in companies' production or shopping stores allow visitors to sample various types of olive oil, learning to distinguish between different flavors, qualities, and grades. These tastings are often accompanied by explanations of how to properly taste and evaluate olive oil [1, 8].
- Workshops concerning olive oil: many olive oil tourism sites offer workshops or classes on topics such as cooking with olive oil, pairing olive oil with food, and understanding its health benefits [3, 13].
- Cultural experiences: olive oil tourism often includes cultural and historical elements, such as visiting museums dedicated to the history of olive cultivation, participating in traditional olive harvest festivals, and learning about the role of olive oil in local cuisine and traditions [14, 15].
- Local Cuisine: many tours include meals or snacks featuring dishes prepared with local olive oil, providing a culinary experience that highlights the versatility and flavor of high-quality olive oil [16], which creates a notable gastronomic experience for the tourists [17].

For this reason, as suggested by Čehić et al. [12] and Almeida and Silveira [11], some attractions are necessary for the development of olive oil tourism. Accordingly, they are olive groves and olive farms, oil mills open for visitors, museums, and interpretation centers dedicated to olives and olive oil, traditional and/or protected geographical indications of production, events/fairs, devoted to olives and olive oil, olive oil roads, and specialized shops for olive oil – oleotece.

However, as in other type of industry, all these oil tourism attractions, innovation, and creativity are seen as one of the principal keys to induce the sustainable development of the rural tourism product [5].

2.2 Innovation trough tradition: A new approach

The concept of innovation can be defined in different manners. However, there is a certain consensus that the approach of innovation can be due to Schumpeter [18], who emphasizes the entrepreneur as a promoter of innovations, *which creatively disrupt the existing order in the market place* ([19], p. 12). The entrepreneur's motivation to innovate is a fundamental ingredient toward the success of the firm. Innovation—as a factor of distinctiveness or exclusivity—will create a superior image in the market, allowing superior profits to the firm. According to Schumpeter [18], these innovations can take the form of (i) the introduction of a new product as well as (ii) the discovery of new source of raw materials, (iii) process innovation, (iv) organizational innovation, and (v) market sourcing. In this same line, the Oslo Manual developed jointly by the OECD and Eurostat (the international reference guide for collecting and using data on innovation) suggests four different types of innovation: product innovation, process innovation, marketing innovation, and organizational innovation [20].

According to the aforementioned, innovation refers not only to the development of new products (with different characteristics and performances), but also to organizational processes (more effective ways to organize), process innovation (new ways to produce products and services), and marketing approaches (new or more effective ways to approach and deal with market and segments). In this sense, the success of any innovation (and consequently the competitiveness of firms/regions) depends on its marketing orientation; that is, on its capability to adapt or anticipate market tendencies.

In an era of valorization of the past, search of new and authentic experiences and scarce material resources of rural areas, they need to valorize their unique and rare endogenous (material or immaterial) resources. These resources are viewed as relics of the past and also as valuable testimonies of past memories and products, reflecting the 'cultural richness' of a destination (i.e., region, city, or country), and consequently reinforce its identity [21, 22]. This is the argument to emphasize ITT.

As suggested by De Massis et al. [23], ITT is based on two key capabilities: interiorization and reinterpretation (**Figure 1**).

Accordingly, interiorization allows assimilating and sharing past knowledge of the firm's tradition territory's tradition across the entire organization, as reflected by the different forms of codified and tacit knowledge used to develop new services or products. Reinterpretation, in turn, allows the combination of selected forms of past knowledge with up-to-date technologies to turn them into new products. Moreover, when firms interiorize past knowledge, this can take different forms, both codified and tacit, that nourish the innovation process [24, 25].

Summing up, ITT is a concept that emphasizes how innovation can emerge from deep engagement with and reinterpretation of traditional practices, knowledge, and values of a society. It suggests that rather than viewing tradition as outdated or stagnant, it can be a rich source of inspiration and innovation. This approach not only honors the past but also ensures that tradition remains a living, dynamic force that continues to evolve and contribute to the future.

Despite the recognition of the importance of ITT, to date, research that focuses on the concept in olive oil tourism is scarce. Bezerra and Correia [3], for instance, argue that olive oil tourism, a product of strong traditions, deeply rooted in the culture of diverse countries, allow visitors to have meaningful experiences, to learn more about the local skills and traditions and to engage with local people and local culture. Although, the authors present diverse experiences of olive oil tourism in the northern region of Portugal and do not explore in a depth manner, the tool and value of tradition in the product. Soares and Dias [26], in turn, claim that the extensive roll of experiences in olive oil tourism can constitute the starting point for the structuring of an offer creative and multidimensional that will contribute to the

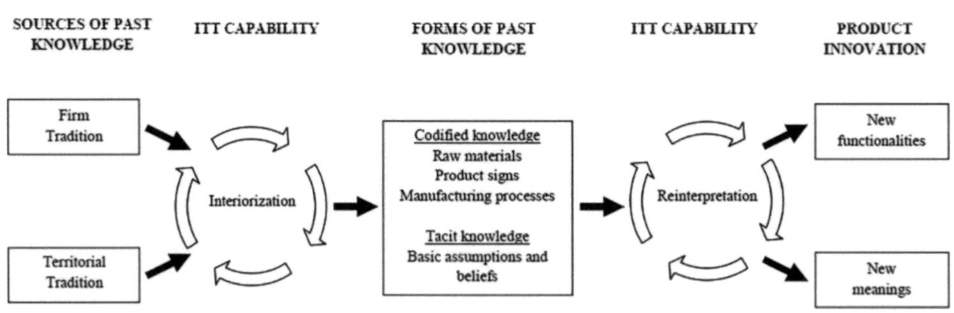

Figure 1.
Model of innovation trough tradition. Source: De Massis et al. [23].

enrichment of niche tourist products in the olive territories and a suitable response to the needs of various tourist segments. But again, these authors do not explore the role of tradition in olive oil tourism. Indubitably, this gap in the literature justifies a more incisive approach to the concept of ITT in the rural product.

3. The region under analysis and methodological procedures

3.1 The territory under analysis

In Portugal, olive oil produced in the country can be protected in six different regions through the PDO region designation. Three of these PDO sub-regions—Norte Alentejano, Moura, and Alentejo Interior—are located in the NUT of Alentejo, in the center-south of mainland Portugal [27]. As with other traditional products from the European Union (EU) and Portugal, the PDO designation adds value to the olive oil and also gives the product recognition for its origin combined with the cultural and ancient tradition of its production [28, 29].

The Alentejo region is also the region with the largest concentration of olive groves (52.4%) [30]. Of the 377,234 hectares in Portugal Mainland, around four-fifths are located precisely in Alentejo, and 70% of the production of olive oil in the country comes from this region [30]. Olive groves are located on plains with arid climates, resulting in smooth but complex olive oils with a fruity flavor, which plays a significant role in the local economy and culture. Indeed, presently, olives are part of the Alentejo's culture, landscape, cuisine, and heritage.

In the region olive oil tourism is becoming increasingly popular, with visitors able to tour olive groves, visit mills, and participate in tastings. Many estates offer tours that include insights into the olive oil production process and the opportunity to sample different oils paired with local foods. Because of this, Alentejo is home to many olive oil producers, ranging from small family-run estates to large, internationally recognized companies. Here are some of the notable olive oil farms open to visitors in the region: Herdade do Esporão, Olimel, Santa Vitória, Courela do Zambujeiro, and AmoreCego. With the exception of one farm, they are family-owned groves, with small, family-run olive farms which often offer personalized tours and insights into traditional methods of cultivation and harvesting.

In summary, olives are a central part of life in Alentejo, both economically and culturally. The region's dedication to quality and tradition makes its olive oil highly respected both in Portugal and internationally.

3.2 Methodological procedures

A qualitative approach was chosen for this study due to the exploratory nature of the work [5] and the scarcity of previous studies in the area. In a first stage in the referred sub-regions, it was done a search in Google engine with the purpose to find olive oil farms that have visits of tourists. For their apparent notable expression in the region, it was selected the following farms: Herdade do Esporão, Olimel, Santa Vitória, Courela do Zambujeiro, and AmoreCego. With the exception of one farm, they are family-owned groves, with small and family-run olive farms.

In a second stage of the process, the websites of these farms are observed and analyzed within the context of searching for any relationships between the use of traditional resources and innovation. Following Presenza et al. [5], findings are grouped in three main categories: the role of tradition; the source of tradition; and the recombinant strategies.

4. Results

The case analysis reveals several topics of particular interest concerning how tradition can inspire innovation in the different activities of the olive oil tourism farms. It seems that through this blend of old and new, olive oil tourism not only preserves heritage but also paves the way for a sustainable and vibrant future. As mentioned before findings are grouped in three main categories: the role of tradition; the source of tradition; and the recombinant strategies.

4.1 The role of tradition in olive oil tourism

Tradition plays a central role in olive oil tourism, serving as a bridge between cultural heritage, agricultural practices, and modern tourism experiences. Olive oil is deeply rooted in the history and culture of diverse agricultural farms in the Alentejo Region, such as the case of all olive oil farms open to visitors mentioned above. The traditional methods of cultivation, harvesting, and production of olive oil are often passed down through generations, forming a core part of the cultural identity of the farms. Indeed, many olive oil producers emphasize traditional methods of cultivation, such as the use of traditional varieties of olive trees and historical olive groves. Some olive groves are hundreds or even thousands of years old, offering a glimpse into the ancient practices of olive cultivation. See for instance the information gathered from AmoreCego, Courela do Zambujeiro, and Olimel websites.

We started the dream of producing olive oil from a small olive grove, taking advantage of an olive oil heritage planted by our grandfather more than six decades ago [31].

With an olive grove with more than 400 centuries-old olive trees, we quickly rediscovered the flavors of our childhoods and awakened in us the desire to fulfill our dream of producing excellent olive oil [32].

It is a 100% natural product obtained thanks to ancient Portuguese olive trees that benefit from a pure Mediterranean climate and fertile soil [33].

Also, the use of traditional techniques of olive oil production or using old-fashioned presses are a reality in some of the mentioned farms. Traditional methods are often associated with higher quality olive oil, and this emphasis on craftsmanship is a key theme in many tours in order to distinguish their product from mass-produced olive oils.

All the farms, visits are allowed in the olive groves and sometimes participate in age-old practices that have remained largely unchanged. During these visits, farms share stories and insights about the history and traditions of olive oil in the region. This authenticity is a major selling point, as tourists seek genuine and immersive experiences.

Tours frequently also include tastings of olive oil paired with traditional local foods, which further embeds the experience within the regional gastronomic culture and culinary traditions (**Figure 2**).

It should be noted that tourists are often educated on how traditional practices contribute to the flavor, aroma, and overall quality of the olive oil. These have the opportunity to purchase traditionally made olive oil, traditional foods, soaps, cosmetics, and other products directly from producers.

Figure 2.
Olive oil tasting offered by "Amor é Cego" farm. Source: https://azeiteamorecego.pt/pt/#visitas.

Last, but not the least, some farms (e.g., AmoreCego, Herdade do Esporão, or Courela do Zambujeiro) also offer accommodation, offering a deeper connection to the local culture and lifestyle.

4.2 The inspiration for tradition in olive oil tourism

In this section, findings are analyzed and used to categorize several sources of tradition that inspire ITT in olive oil tourism. The result is the grouping of the sources of tradition into three main types: (a) tradition derived from the olive oil farm itself; (b) tradition from an ancient production; and (c) tradition from the olive oil territory and destination.

4.2.1 Tradition derived from the olive oil farm itself

Most of the firms of olive oil production in Portugal and consequently in Alentejo Region are small family-run firms [34], where one family or owner manages the property and business. From the owner perspective, the olive farm is often viewed as "*the feeling that moves us, it is the passion we have for our land, in the family for three generations and which with a lot of effort and affection we have preserved* [31]. Among others, this is also the case of Courela do Zambujeiro firm:

A delight when the winter ends, when the ground covers of multicolored flowers, where we can see rabbits walking between trees, storks returning at the end ofr noon, where we can hear sheeps bleat out, at night observe the owl which looks at us on its branch of olive tree, smell all the aromas which wrap Farm and which wake in us the sensations of a world which we believed has for ever disappeared [32].

4.2.2 Tradition from a past production

The production of olive oil is one of the oldest agricultural traditions, with its roots tracing back thousands of years to ancient civilizations in the Mediterranean region [35]. In many ancient Mediterranean cultures, olive oil was not just a staple of the diet but also held religious and symbolic significance. It was used in religious rituals, for anointing, and even in the sacred lamps in temples [36]. In Ancient Greece, olive oil was considered a gift from the Gods, and in Rome, it was used for everything from cooking to bathing, athletic training, and medicinal purposes [36]. Additionally, the olive tree is observed as a symbol of peace, value, spirituality, and sacredness – an element of strength and purification, of resistance to the ravages of time and wars [37].

Additionally, yet it must be considered the importance of secular olive trees not only from a cultural, environmental and scientific point of view, but also due to the precious genetic heritage of which they are depositaries [15]. This idea is evident in the following farm statement:

From the work of three generations who come together around the same objective, the result is products that resemble traditional Portuguese flavors and their recognized medicinal properties. By bringing these foods back to your table, you can associate the memory of a healthy past with an even healthier future [33].

4.2.3 Tradition from the destination

The EU agricultural product quality policy, introduced in 1992 (Council Regulation (EEC) N. 2081/92 of 14 July), on the protection of geographical indications and designations of origin schemes and presently governed by the Council Regulation (EU) 1151/2012 of 21 November, aims to emphasize the quality of products resulting from a specific origin and/or its inherent natural and human factors [38]. This document stresses the definitions of protected designations of origin (PDO) and protected geographical indications (PGI) for agricultural products and foodstuffs, also called traditional agri-food products [27]. As state by Pato and Duque [29] a strong link with territory, that is, the destination, is a condition of this type of protection. Indeed, *the distinctive character based on the territory and their culture creates and adds value to these products, and also gives the product recognition for its origin combined with the cultural tradition of the place* ([29], p. 2).

Portugal has an extensive and diverse range of olive oils from traditional nature, associated with each of the regions of the country and to Mediterranean diet, which resulted from the cultural and historical influence in the elaboration of these foods, the living heritage of a unique and rich gastronomic patrimony.

As mentioned above tree of the olive oil PDO regions of production are located in Alentejo region—Norte Alentejano, Moura, and Alentejo Interior [27]. In these regions are produced *Azeite do Norte Alentejano* PDO (made particularly by Galega olive variety), *Azeite do Alentejo Interior* PDO (produced by mechanical processes with local varieties, particularly Galega Vulgar olive variety (60%), Cordovil de Serpa and/or Cobrançosa), and *Azeite de Moura* PDO (obtained by mechanical processes, particularly from olives of the Galega, Verdeal and Cordovil varieties), respectively [27].

4.3 Intertwine strategies

In a typical production in the region, as suggested by Presenza et al. [5], the most creative and competitively advantageous way of using tradition to strategically innovate is through a combination of diverse forms of tradition. This often requires the olive oil suppliers to revive partial (or sometimes complete) the values of the past, to reproduce the experience for guests. Indubitably, this involves thinking about the environment in which the ancient and cultural elements originate and reviving those activities and efforts necessary to create the main characteristics that authenticate the experience. Of course, some adaptation to the modern world will be necessary. For instance, those that are related to sanitary, hygienic, and security of the experience. But in all the cases mentioned above, there is no doubt that safety conditions are assured. Another requisite, it is the comfort that tourists also want. The interdependencies and the opportunity (and often the need) to combine traditional resources with the "modern world" are very clear in all the cases presented here. The purpose, indeed, is to integrate and link together the various components of traditional and more modern resources in order to create a coherent and integrated offer and experience.

In fact, the reality that strongly emerges is that olive oil tourism is not only singularly focused on the revitalization of existing traditions but also integrate these traditions with the modern world in a balanced way. An example of this is proposed by the Restaurant Herdade do Esporão:

> *The cuisine offers a modern interpretation and preparation, centred around the product – 'Alentejano', Portuguese, seasonal. The ingredients are grown in our organic vegetable gardens or come to us from local and national producers we work with all year round. We respect the product throughout its entire cycle, maximising it to the fullest. Our kitchen is sustainable and zero waste [39] (***Figure 3***).*

Also, nowadays, the farms offer a lot of commodities, such as a foyer with Wi-Fi access. It is also interesting to note that some businesses offers a lot of different experiences. and activities. Santa Vitória firm, for instance offers the following activities/experiences: olive picking, wine tasting, picnic, carriage ride, equestrian activities,

Figure 3.
Restaurant of Herdade do Esporão. Source: https://guide.michelin.com/pt/pt_PT/article/features/migas-alentejanas-a-gastronomia-sustentavel-que-aproveita-os-sabores-e-encanta-cada-vez-mais

Figure 4.
Olive oil meal offered by "Amor é Cego" farm. Source: https://azeiteamorecego.pt/pt/#visitas.

hot air balloon ride, bird watching, pedal karts, bicycle rides, jeep ride, paintball, family go-kart, archery, canoeing, educational farm, and kitchen workshop, among others [40]. Also, besides the offer of different activities, the owner of Amor é Cego firm has also developed a new project "Oliveira Velha", born from the friendship between "Amor é Cego" and "Tua Madre" restaurant. It is an experience developed in the nature through which visitors can be in the middle of nature, walk through the olive grove, learn more about the production of olive oil and its benefits, hold a technical olive oil tasting, and finally, at the table, they will have a gastronomic experience based on local products and in the Galega olive oil (**Figure 4**).

Undoubtedly, it is a way to combine the past and the present and also a way to combine different activities and offer a more complete touristic product to the tourists. Moreover, the mixture of experiences and strategies associated with innovation in the positioning of the farm offer clearly go beyond the business and its owners. In fact, the development of synergies between firms and other suppliers in the rural community *can be seen as a defining characteristic that often involves the entire destination in producing a unique combination of identity and local entrepreneurial skills* ([5], p. 72).

5. Conclusions

The paper highlights how ITT is gaining attention as a critical issue for rural tourism, particularly in olive oil tourism. A review concerning the thematic suggests using traditional resources and capabilities of rural areas as a mean to deal in an industry very competitive [5, 6, 23]. The growing interest of tourists in this area could be translated into strategic advantage by the tourism in rural areas and be of interest to researchers [5].

Indeed, rural tourism entrepreneurs can be inspired by traditional resources to innovate in diverse fields. This emphasizes that, the basis of every organization's sustained competitive advantage rests on its heterogeneous resources and/or capabilities, which is claimed by the resource-based view (RBV) of the firm. This theory suggests that sustained competitive advantage for an organization depends, above anything else, on a roll of resources, which are of rare value, heterogeneous in nature, and, consequently, not easily copied by competitors or perfectly transferable by them [41, 42]. The use of tradition, as a rare, valuable, inimitable, and non-substitutable resource, influences innovation relating to products/services, processes, and organization by the olive oil firm. For instance, food experiences in the farm with all the commodities are a form of product innovation; the process of obtaining olive oil using traditional and ancient techniques of production are a form of process innovation, while the links between other entities and suppliers are a form of organizational innovation.

Moreover, on the one hand the findings present here reveal three main types of sources of ITT: i) the firm itself, ii) knowledge related to a past rural activity, and iii) tradition from the rural destination. Hence, as suggested by Presenza et al. [5], this reinvention depends on the manager's/owner's ability to reinterpret the traditional resources of the farm/region. On the other hand, the offer of olive oil farms illustrates how entrepreneurs create value for tourists/customers through the revitalization, preservation, and strategic reinvention of tangible and intangible resources of a specific tradition.

The findings of this study lead to conclude that ITT in olive oil tourism has the following benefits:

- Cultural continuity: ITT helps maintain a sense of cultural identity while adapting to changing times.

- Sustainability: traditional practices are often inherently sustainable and can be adapted to address modern environmental challenges.

- Differentiation: in a globalized world, leveraging tradition can provide a unique value proposition, differentiating products, services, or experiences.

- Resilience: By blending the old with the new, societies and organizations can build resilience, drawing on the strength of tradition while being flexible enough to innovate.

Summing up, tradition in olive oil tourism is essential for offering an authentic, immersive experience that connects visitors to the history, culture, and craftsmanship of olive oil production. It fosters a deeper appreciation of the product and the region, supports local economies, and helps preserve agricultural and culinary traditions that might otherwise be lost. Because that, olive oil tourism is also an innovative and also a traditional form of rural tourism.

The use of tradition in olive oil tourism has several important implications for olive oil tourism providers, policymakers, and destination marketers. Tradition offers a rich cultural narrative that can enhance the overall experience and value of olive oil tourism and contribute to the development of the olive oil territory. For providers, it adds value to the tourist experience through authenticity and education. Policymakers can use it as a tool for cultural preservation and sustainable development, while destination marketers can build strong brand identities and attract niche markets by promoting the rich cultural heritage of olive oil production.

This study has naturally some limitations. While the qualitative analysis of websites is a valuable approach, it would be useful to complement the data collected with other additional data collection methods. Namely, interviews with olive oil tourism providers or surveys of tourists would be important in order to strengthen the validity and reliability of the findings. At the same time, it would permit to observe some additional practices in course. Employing triangulation techniques, which involve using multiple data sources and methods to verify findings, could actually enhance the credibility of the research.

Secondly, because of time constraints, it would be also interesting to extend the study to other regions and/or entities (such as the public ones). This integration leads to have a more complete understanding of the total offer in this recent and amazing sector of activity.

Acknowledgements

This work was funded by National Funds through the FCT – Foundation for Science and Technology, I.P., within the scope of the project Refª UIDB/00681/2020. Furthermore, we would like to thank the CERNAS Research Centre and the Polytechnic Institute of Viseu for their support.

Conflict of interest

The authors declare no conflict of interest.

Author details

Maria Lúcia Pato
CERNAS-IPV Research Centre, Polytechnic Institute of Viseu, Campus Politécnico, Viseu, Portugal

*Address all correspondence to: mluciapato@gmail.com

References

[1] Pato ML. A decade of olive oil tourism: A bibliometric survey. Sustainability. 2024;**16**(4):1665

[2] Pulido-Fernández JI, Casado-Montilla J, Carrillo-Hidalgo I. Analysis of the Demand Behavior of the Olive Oil Tourism from the Supply Perspective. Alicante: Universidade de Alicante; 2021

[3] Bezerra R, Correia AI. The potential of olive oil for creative tourism experiences in the northern region of Portugal. Revista Portuguesa de Estudos Regionais. 2019;**51**:55-72

[4] Hjalager A-M, Kwiatkowski G, Østervig Larsen M. Innovation gaps in Scandinavian rural tourism. Scandinavian Journal of Hospitality and Tourism. 2018;**18**(1):1-17. DOI: 10.1080/15022250.2017.1287002

[5] Presenza A, Messeni Petruzzelli A, Sheehan L. Innovation through tradition in hospitality. The Italian case of Albergo Diffuso. Tourism Management. 2019;**72**:192-201. DOI: 10.1016/j.tourman.2018.11.020

[6] Messeni Petruzzelli A, Savino T. Reinterpreting tradition to innovate: The case of Italian haute cuisine. Industry and Innovation. 2015;**22**(8):677-702

[7] Murgado-Armenteros EM, Parrilla-González JA, Medina-Viruel MJ. What does the olive oil tourist value at the destination? A criterion for olive oil tourism segmentation. International Journal of Gastronomy and Food Science. 2021;**25**:100378. DOI: 10.1016/j.ijgfs.2021.100378

[8] Hernández-Mogollón JM, Di-Clemente E, Campón-Cerro AM, Folgado-Fernández JA. Olive oil tourism in the Euro-Mediterranean area. International Journal of Euro-Mediterranean Studies. 2021;**14**(1):85-101

[9] Karaman S, Budak F. Evaluation of Olive Plant in the context of faith tourism. Request & Demand Address. 2023;**126**:126-133

[10] Pulido-Fernández JI, Casado-Montilla J, Carrillo-Hidalgo I. Introducing olive-oil tourism as a special interest tourism. Heliyon. 2019;**5**(12):e02975

[11] Almeida I, Silveira L. Olive-oil tourism in Portugal. Revista Turismo & Desenvolvimento. 2021;**36**(2):153-165. DOI: 10.34624/rtd.v36i2.23289

[12] Čehić A, Mesić Ž, Oplanić M. Requirements for development of olive tourism: The case of Croatia. Tourism and Hospitality Management. 2020;**26**(1):1-14

[13] Murgado-Armenteros EM. Turning food into a gastronomic experience: Olive oil tourism. Options Mediterranéennes. 2013;**106**:97-109

[14] Şahin Ö, Aydin A. A cultural, gastronomic, and touristic asset: The Kuşadası oleatrium olive and olive oil history museum case. Journal of Tourism & Gastronomy Studies. 2017;**5**(4):180-199

[15] Schicchi R, Speciale C, Amato F, Bazan G, Di Noto G, Marino P, et al. The monumental olive trees as biocultural heritage of Mediterranean landscapes: The case study of Sicily. Sustainability. 2021;**13**(12):6767

[16] Dancausa-Millan G, Millán-Vázquez de la Torre MG, Huete-Alcocer N.

Olive oil as a gourmet ingredient in contemporary cuisine. A gastronomic tourism proposal. International Journal of Gastronomy and Food Science. 2022;**29**:100548. DOI: 10.1016/j.ijgfs.2022.100548

[17] Björk P, Kauppinen-Räisänen H. Interested in eating and drinking? How food affects travel satisfaction and the overall holiday experience. Scandinavian Journal of Hospitality and Tourism. 2017;**17**(1):9-26. DOI: 10.1080/15022250.2016.1215871

[18] Schumpeter J. Theory of Economic Development: An Inquiry into Profits, Capital, Credit, Interest and the Business Cycle. Cambridge: Havard University Press; 1934

[19] Dinis A. Marketing and innovation: Useful tools for competitiveness in rural and peripheral areas. European Planning Studies. 2006;**14**(1):9-22. DOI: 10.1080/09654310500339083

[20] OECD, & Eurostat. OSLO Manual Guidelines for Collecting and Interpreting Innovation Data. Paris: OECD; 2005

[21] de Saint-Laurent C, Obradović S. Uses of the past: History as a resource for the present. Integrative Psychological and Behavioral Science. 2019;**53**:1-13

[22] Paniccia P, Pechlaner H, Valeri M. The importance of the time of experience in the innovation of tourism business: The sextantio Albergo Diffuso. In: Entrepreneurship and Innovation in Tourism. Berlin: Erich Schmidt; 2010. pp. 97-116

[23] De Massis A, Frattini F, Kotlar J, Petruzzelli AM, Wright M. Innovation through tradition: Lessons from innovative family businesses and directions for future research. Academy of Management Perspectives. 2016;**30**(1):93-116

[24] Haruyama T. Competitive innovation with codified and tacit knowledge. Scottish Journal of Political Economy. 2009;**56**(4):390-414

[25] Senker J. The contribution of tacit knowledge to innovation. In: Cognition, Communication and Interaction: Transdisciplinary Perspectives on Interactive Technology. London: Springer Science & Business Media; 2008. pp. 376-392

[26] Soares I, Dias F. The regeneration of the olive heritage in Portugal supported by creative tourism. Revista Portuguesa de Estudos Regionais. 2024;**68**:29-46. DOI: 10.59072/RPER.VI68.603

[27] DGADR. Produtos Tradicionais Portugueses. 2024. Available from: https://tradicional.dgadr.gov.pt/pt/cat/azeites-e-azeitonas

[28] Mattas K, Baourakis G, Tsakiridou E, Hedoui MA, Hosni H. PDO olive oil products: A powerful tool for farmers and rural areas. Journal of International Food & Agribusiness Marketing. 2020;**32**(4):313-336. DOI: 10.1080/08974438.2019.1599763

[29] Pato ML, Duque AS. Traditional Agri-food products and sustainability–a fruitful relationship for the development of rural areas in Portugal. Open Agriculture. 2023;**8**(1):20220157

[30] INE. Recenseamento Agrícola 2019. Lisboa: INE; 2021

[31] AzeiteAmoreCego. Amor é Cego. 2015. Available from: https://azeiteamorecego.pt/en/

[32] CoureladoZambujeiro. Courela do Zambujeiro. 2015. Available from:

https://www.coureladozambujeiro.com/a-courela/

[33] Olimel. Olimel - Azeite e Mel. 2022. Available from: https://olimel.pt/

[34] GPP. Análise Setorial do Azeite. Lisboa: GPP; 2020

[35] Efe R, Soykan A, Cürebal I, SönmezIn S. Olive and Olive Oil Culture in the Mediterranean Basin. In: Efe R, Ozturk M, Ghazanfar S, editors. Environment and Ecology in the Mediterranean Region. New Castle: Cambridge Scholars Publishing; 2012. pp. 51-62

[36] Kostelenos G, Kiritsakis A. Olive tree history and evolution. In: Olives and Olive Oil as Functional Foods: Bioactivity, Chemistry and Processing. Oxford: John Wiley & Sons; 2017. pp. 1-12

[37] Grego S. The olive tree: A symbol. In: Yonar T, editor. Olive Cultivation. London, UK: IntechOpen; 2022

[38] FAO. The Nutrition and Health Potential of Geographical Indication Foods. Rome: FAO; 2021

[39] HerdadedoEsporão. (s.d.). Herdade do esporão Restaurant.Available from: https://esporao.com/en/tourism/herdade-do-esporao-restaurant

[40] SantaVitória. Activities and Experiences. 2024. Available from: https://www.santavitoria.pt/pages/atividades-experiencias

[41] Barney J. Firm resources and sustained competitive advantage. Journal of Management. 1991;**17**(1):99-120

[42] Kruesi MA, Bazelmans L. Resources, capabilities and competencies: A review of empirical hospitality and tourism research founded on the resource-based view of the firm. Journal of Hospitality and Tourism Insights. 2023;**6**(2):549-574. DOI: 10.1108/JHTI-10-2021-0270

Chapter 7

Mediterranean Treasures: Olive Varieties for Table and Olive Oil

Perihan Adun and Şebnem Güler

Abstract

This study delves into the diverse varieties of table olives and olive oils in the Mediterranean Basin, renowned for its rich cultural heritage and culinary traditions. Focusing on the agricultural significance and gastronomic versatility of these products, the research explores varieties, their unique flavors, nutritional profiles, and historical importance. By examining their regional distinctions and health benefits, the study aims to highlight the integral role of olives and olive oil in Mediterranean cuisine and lifestyle. Ultimately, this chapter aims to emphasize the importance of managing cultivated olive germplasm sustainably for future breeding programs. Additionally, preserving wild and feral olive populations is crucial in the context of global warming and climate change.

Keywords: olive tree, genealogical tracing, olive oil, table olives, varieties, functional foods, suitability, drought, climate change

1. Introduction

Olive tree is a sacred tree in ancient literature and shall be regarded as a sacred tree even in our times since it untimely represents life, abundance, resilience, health, beauty, hope, and peace, taking into consideration its unique and excellent characteristics like longevity, beauty of branches and flowers, and high nutritional value of fruits, leaves, and oil (**Figure 1**) [1]. The archaeobotanical records agree that the olive tree is one of the oldest trees together with grape, fig, date, pomegranate, and almond in the Mediterranean Basin, from the Near East to the western Mediterranean, going backward more than 6000 years in human history [2–4].

Olive oil had many uses in ancient times beyond cooking and eating. It was employed as a perfume for anointing the deceased, making soap, and fueling lamps [5]. In ancient Greece, athletes would ceremonially cover their bodies with it. It has long been a symbol of wealth and influence, used to anoint the heads of esteemed individuals throughout history. Olive branches and crowns were also presented to powerful leaders (**Figure 2**). Olive oil played a role in the production of medicines and cosmetics. Hippocrates referred to it as “the great healer,” while Homer glorified it as “liquid gold.” Galen also paid tribute to its health benefits, and in ancient Greece, it was regarded as a luxury item [6, 7].

Figure 1.
"and the dove came back to him in the evening, and there in its beak was a freshly plucked olive leaf; so Noah knew that the waters had subsided from the earth." Genesis 8: 8–12 [1].

Figure 2.
Woman and Hercules with garland of olive leaves, in Herculaneum fresco (National Archaeological Museum, Naples) ©Carole Raddato.

DOI: http://dx.doi.org/10.5772/intechopen.1008263

The cultivation of olive trees is closely tied to the climate, agricultural landscape, and way of life in the Mediterranean Basin [8]. Wild olive trees, the ancestors of today's domesticated varieties, still grow in regions such as the Peloponnese, Crete, North Africa, and the Middle East, where they first originated (**Figure 3**). The deep connection between the olive tree and human civilization has created a vast cultural heritage, deeply woven into the daily practices of Mediterranean people, from food and art to traditions, with the pivotal role of olive oil [9].

> *"The Mediterranean ends where the olive tree no longer grows.*
> *There were the sun permits, the olive tree takes root and gains ground.*
> *The tree of light is the nature and culture of the Mediterranean".*
> *Georg Duhamel [10].*

During the Mesolithic period (10000–8000 BC), humans crafted small tools from stone, bone, or wood, using them for spears and arrows [11]. During this time agriculture emerged, leading to the development of permanent village settlements, such as those found at the Pre-Pottery Neolithic site of Göbekli Tepe, Türkiye (**Figures 4** and **5**). This hilltop sanctuary was in an open steppe environment, rich in wild cereals like einkorn wheat and barley. Plant remains discovered at Göbekli Tepe include charcoal from *Pistacia atlantica*, *Prunus amygdalus*, *Quercus brantii*, *Populus euphratica*, and *Fraxinus augustifolia* [12–15]. By the Neolithic period (approximately 8000 BC to 3000 BC), humans had shifted from a hunter-gatherer lifestyle to one focused on agriculture and food production [16]. Although it is difficult to determine

Figure 3.
The oldest olive tree in the world grown in the Ano Vouves village of Kissamos in Chania, Crete, as determined by the international scientific community. The ancient tree is 3000 years old and still producing high-quality olive. ©Dimitra Damian/Greek Reporter.

Figure 4.
Göbekli Tepe: Overhead view of the excavation areas ©Erhan Küçük (DAI Orient Department).

Figure 5.
Human history timeline to facilitate reading.

the initial stage of fruit crop domestication, horticulture developed only after grain agriculture in southwest Asia. The most important crops in first farming communities were emmer wheat, einkorn wheat, barley, lentil, pea, and flax in the Levantine Corridor covering Fertile Crescent [17, 18]. Based on evidence from living plants, archaeological findings, and genetic studies, it is usually accepted that the domestication of the olive tree began in the Near East approximately 6000 years ago in Copper (Chalcolithic) and Bronze Age cultures [4, 19–21].

In antiquity, olive cultivation spread to new areas around the Mediterranean Basin and beyond. The trade of olive products was well recorded. The main route of expansion was toward the west. Numerous archaeological sites in Spain, Greece, Egypt, Western Turkey, Jordan, and Palestine have uncovered various artifacts related to olive and olive oil production, such as milling stones, decantation basins, storage containers, frescoes, and ancient texts (**Figure 6**) [22, 23].

These archaeological discoveries indicate that olive cultivation spread across Syria, Israel, and Crete between 5000 and 1400 BCE, coinciding with the expansion of Greek colonies. From there, it reached Southern Italy and Northern Africa, eventually extending to Southern France. Olive trees were cultivated throughout the Mediterranean Basin during the Ancient Roman [24, 25].

Recent evidence has shown that by the Middle Copper Age (around 4600 BCE), pickled and dry-salted table olives were produced in Israel, specifically at the site of Hishuley Carmel [26]. Additional traces of olive cultivation and oil production have been uncovered in Minoan Crete, notably at Aphrodite's Kephali (Early Minoan I, ca. 3200–2700 BCE) [27]. The agricultural practices in Bronze Age Crete do not entirely align with the conventional view that vineyards and olive groves were systematically utilized from the early Bronze Age, forming key components of the Minoan agricultural economy. Systematic wine production likely began during the first Palace period, peaking in the second Palace period, while large-scale oil production appears in the second Palace period and increased during the 'post-palatial' era [28]. Oil was used on many different occasions: for religious rituals and funerals, for lighting and personal cleanliness, yet also as a luxury item to be exchanged or presented. The daily consumption of olive oil in rural Crete was around 80 grams *per capita* and may contribute almost 30% of dietary intake [29].

The Phoenicians played a key role in spreading the olive tree to western regions through trade with other marine centers. By the eighth century BC (800–701 BC),

Figure 6.
2600–2000 BC of Bronze age stone blades used as olive-pruning saws and olive press cut into limestone bedrock. See the circular channel draining the oil from the pressing tank (top) to the collection vat (below) at Khirbet Um al-Ghozlan. North Jordan. ©Adam Carr.

olives began reaching the Greek Islands, Algeria, and Carthage. The Greeks, by founding colonies in other parts of the Mediterranean, such as Spain, introduced olive cultivation to these new regions. The first notable advancements in olive farming took place in the 8th and 7th centuries BC, when cultivation practices became more systematic. Even though grafting was a heavily employed technique by the ancient Greeks and Romans, it resulted in a rapid decline in genetic diversity by using only a few genotypes for the sake of more efficient production [30–32].

The Romans used olive oil primarily in their baths and as fuel, since they did not consider it of high enough quality for consumption [31]. However, with the rise of the Roman Empire and the conquest of Greece, Asia Minor, and Egypt, olive oil gained importance due to expanded trade routes across the Mediterranean Basin. It became a vital commodity, not only as a dietary staple but also for medicinal and energy purposes. Olive cultivation experienced a resurgence in the fifth century AD as marine states launched into thrive. In the fifteenth century AD, missionaries and early settlers introduced vines and olive trees to America. While vines spread widely, olive cultivation was limited to areas like Chile, Argentina, and California, which share Mediterranean-like climates [33].

The expansion of olive tree cultivation was accompanied by advancements in olive oil extraction methods. As early as 5000 BC, people gathered olives and crushed them using stone mortars [24]. The olive paste would flow into a small pot through an inclined stone basin, and by adding hot water, the lighter olive oil would rise to the surface, where it was skimmed off and stored in clay vessels. Initially, olive oil production was limited to fulfilling family needs, with small processing facilities located within households. This domestic production model has been well documented during the Bronze Age in Crete [34–37]. There were two main types of storage vessels, depending on where and what to store: pithos and amphora. Both were made of clay to keep the cost low and take advantage of clay's ability to keep out heat, water, dirt, and pests. A pithos was a storage vessel for liquids like wine and olive, grains, and other types of food as well. A pithos is about a man's height (this one is 1.7 m) and its capacity is around a ton (**Figure 7**). The pithoi were designed mostly for permanent installment in the household. Pithoi could be dug into the ground to keep items cool below the surface and to have easier access like the one in **Figure 7** found in Herculaneum (the Archaeological Park of Herculaneum of ancient Romans town, Italy).

The commonest storage vessel was the amphora, as the pithoi were expensive and in some cases an indication of wealth, i.e., the more pithoi are found in a house, the wealthier it is. The storage amphoras were plain clay vessels without or with very simple decoration with a pointy bottom. This pointy bottom allowed them to fit better into storage (**Figure 8**).

The cultivated olive has recently been introduced to new regions such as the Americas, Mexico, Brazil, Argentina, Peru, China, India, New Zealand, and Australia. This progress is largely due to the growing awareness of the health benefits of olive oil [22, 38].

This chapter explores the historical evolution of the olive, the varieties grown worldwide, and its high nutritional and socioeconomic significance. Olive and olive oil have become key components of Mediterranean cuisine, and we examine the conditions necessary for sustainable olive production in light of historical, archaeological, botanical, and molecular data. Our aim is to highlight the valuable contribution of olive products to human health and emphasize the importance of managing cultivated olive germplasm sustainably for future breeding programs. Additionally,

DOI: http://dx.doi.org/10.5772/intechopen.1008263

Figure 7.
Iron-age Pithos from Crete. It is about 1.6 m tall, and the full pithos weight is close to 2 tons. Images from Wikipedia and Archeoastronomia in Italia.

Figure 8.
Reconstruction of how amphorae might have been stacked on a galley. Girne (Kyrenia) Castle, Girne, North Cyprus.

preserving wild and feral olive populations, which are hybrids of cultivated trees and oleasters, is crucial in the context of global warming and climate change.

The end of the Paleolithic Age marked the close of the last Ice Age, which led to the extinction of many large mammals, rising sea levels, and climate changes that prompted human migration [39]. Archaeological evidence suggests that beginning in the Holocene epoch (11,700 years ago to the present), plant and animal domestication developed independently in different regions, likely due to climate change and population growth. This gradual shift from a hunter-gatherer lifestyle to agriculture involved the selection of crops and animals for domestication, with a focus on desirable traits. The advent of agriculture was a pivotal moment in human history and evolution [40].

From past climate changes to the current one, we are now witnessing significant warming and reduced precipitation in the Mediterranean Basin, which could have serious ecological, social, and economic impacts. The 2023 IPCC report highlights that "Human influence has unequivocally warmed the atmosphere, oceans, and land, largely due to over a century of greenhouse gas (GHG) emissions resulting from energy use, land-use changes, and consumption patterns. Between 2011 and 2020, the global surface temperature was about 1.1°C higher than in 1850–1900 (1.09 [0.95–1.20]°C), with land temperatures rising faster (1.59 [1.34–1.83]°C) than ocean temperatures (0.88 [0.68–1.01]°C). The atmosphere, oceans, cryosphere, and biosphere have all experienced widespread and rapid changes. Evidence of extreme weather events such as heatwaves, heavy rainfall, droughts, and tropical cyclones, and their attribution to human activity, has strengthened across all regions. Climate change has also contributed to desertification and worsened land degradation, particularly in coastal areas, river deltas, drylands, and permafrost regions (high confidence). Global warming is expected to continue in the near future (2021–2040) due to the continued rise in cumulative CO2 emissions. Every region of the world is projected to experience increased climate hazards, putting ecosystems and human populations at greater risk" (**Figure 9**) [41].

Because of differences in ability of olive varieties to adapt specific soil and climatic conditions can affect their growth and hence oil yield and quality [42, 43]. Politicians, land managers, and oil growers should be encouraged to develop adaptation strategies, even though the olive is resilient to the warm environmental conditions of the Mediterranean Basin [39, 44].

"Measures on the demand side should focus on transitioning to sustainable, healthy diets, minimizing food loss and waste, and promoting sustainable agricultural intensification. By doing so, it is possible to lower CH_4 and N_2O emissions, prevent ecosystem conversion, and encourage land reforestation and ecosystem restoration" [41].

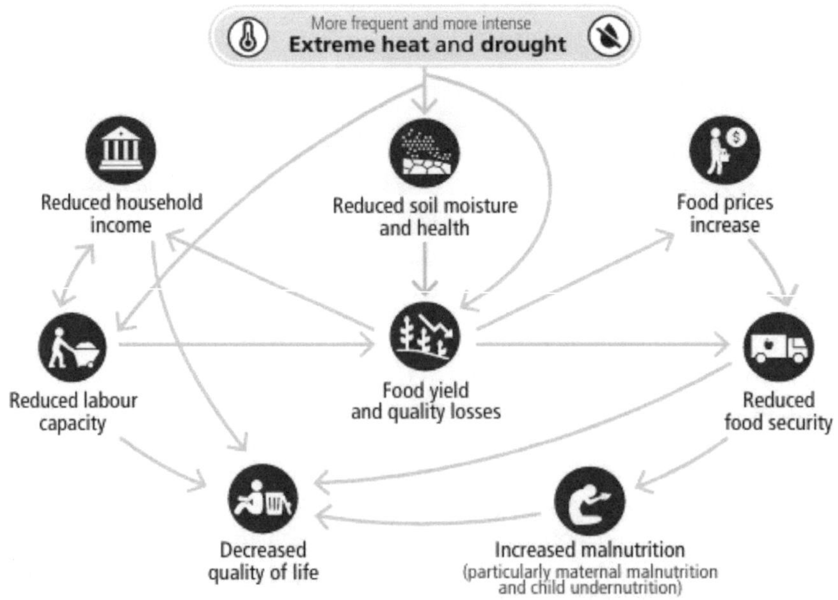

Figure 9.
Multiple climate change risks in the near term (IPCC-2023 Report).

2. Taxonomy and morphology

2.1 Taxonomy

The olive tree belongs to the 29 genera of the Oleaceae family of dicotyledons, which also contains various important ornamental plants such as *Forsythia, Fraxinus, Jasminum*, *Ligustrum*, and *Syringa* (**Table 1**) [45].

The genus Olea was previously thought to encompass around 40 species and subspecies [46]. However, recent studies have shown that this genus is polyphyletic, and it should now be confined to only seventeen species or subspecies belonging to three lineages (E1, E2, and E3) that still exist across Africa, southern Europe, Asia, and Oceania [20]. Among these, the E1 lineage is the most widely distributed worldwide. The olive cultivars associated with the E2 and E3 lineages primarily originated from the central western Mediterranean Basin (CWMB). Early growers selected the "first olive cultivars" based on desirable traits, choosing one from the var. *sylvestris* in the Eastern Mediterranean Basin (EMB) within the E1 lineage and another from the var. *sylvestris* in the E2 and E3 lineages in the CWMB. Subsequently, the process of improvement and selection involved introducing non-native cultivars primarily from the EMB, which led to the replacement of the original "western" olive trees through both crossing and grafting techniques.

Within this family, the olive belongs to the genus *Olea* and the species *europaea*, which has six subspecies: *europaea* (the Mediterranean olive tree), *guanchica, cerasiformis, maroccana, laperrinei*, and *cuspidata*. These subspecies are distributed across the Old World, including the Macaronesian islands. The Mediterranean subspecies encompasses the oleaster, a wild form (*Olea europaea* subspecies *europaea* variety *sylvestris*), and the cultivated olive (*Olea europaea* subspecies *europaea* variety *europaea*) [47]. The O. *europaea* complex has undergone various processes, including hybridization, polyploidy, and geographical isolation, leading to the formation of seven independent lineages, each exhibiting specific morphological traits categorized into subspecies.

Eukaryotes (nucleated cells)	
Kingdom	**Green plants**
Subkingdom	Tracheobionata - vascular plants
Superdivision	Spermatophyta - seed plants
Division	Magnoliophyta - flowering plants
Class	Magnoliopsida - Dicotyledons
Subclass	Asteridae
Order	Scrophulariales or Lamiales
Family	*Oleaceae*
Genus	*Olea*
Species	*europaea*
Variety	*Olea europaea sativa* (cultivated olive seedlings: olivasters)
	Olea europaea sylvester (Mediterranean wild olive seedlings: oleasters)

Table 1.
Taxonomic classification of olive trees.

Of the six subspecies, only three are found naturally in Europe: subsp. *europaea*, which is located in the Mediterranean Basin (including Spain, Italy, Turkey, Greece, Portugal, France, Cyprus, Slovenia, and Malta), and some Atlantic enclaves in southwestern Europe. *O. europaea* subsp. *guanchica, cerasiformis*, and *maroccana* are endemic to the Macaronesian region, with subsp. *guanchica* found in the Canary Islands and subsp. *cerasiformis* in the Madeira archipelago. *Guanchica* is closely related to subsp. *europaea* and has contributed to the formation of the allotetraploid subsp. *cerasiformis*. The allohexaploid subsp. *maroccana* is confined to Morocco, while subsp. *laperrinei* is now limited to the high mountains of the Sahara Desert. In evolutionary dendrograms, subsp. *laperrinei* is closely associated with the E1 lineage of *O. europaea* subsp. *europaea*. Phylogenomic analyses also support the recognition of an additional taxonomic subspecies, *ferruginea*, for Asian populations, which is distinct from the African wild subsp. *cuspidate*. *O. europaea* subsp. *cuspidata* is widely distributed across various native agro-ecological zones, ranging from areas without cultivated forms, such as East Africa, including Ethiopia and Kenya, to the regions where olive cultivation has been recently introduced, like South Africa, India, China, and Nepal [18, 21, 32, 46, 48].

The olive tree features of numerous varieties display both significant and subtle phenotypical and genetic variations. Currently, many of these differences in size, color, oil content, fatty acid composition, and other characteristics have been documented in the primary olive-producing nations [49]. Among the olive cultivars of *O. europaea* subsp. *europaea* var. *europaea*, eleven distinct chlorotypes have been identified. Six of these belong to lineage E1 (E1.1, E1.2, E1.3, E1.N5, E1.N6, and E1.N7), four to lineage E2 (E2.1, E2.3, E2.N4, and E2.N5), and one (E3.1) to lineage E3. The most prevalent profile is E1.1, accounting for 80% of the findings across all studied regions [32].

Both wild and cultivated olives are diploid ($2n = 2x = 46$) and are primarily allogamous. Wild olives reproduce sexually via wind pollination, and their seeds are spread by birds. Cultivated varieties have originated from wild olives through cultural methods like cuttings or grafting [38, 50–53].

Genetic diversity is greater in oleasters than in cultivated olives. The most recent common ancestors of each Mediterranean lineage can be traced back to the Middle Pleistocene: approximately 139,100 years ago (BP) for lineage E1, 284,300 BP for E2, and 143,700 BP for E3. These molecular dating analyses suggest that oleaster populations diversified well before the Last Glacial Maximum (LGM), which occurred between 26,500 and 19,000 BP. Additionally, it appears that the E2 haplotypes specific to the central Mediterranean may have emerged during the colonization after the LGM [4, 54].

A comprehensive genetic survey involving hundreds of olive genotypes and a broad array of chloroplast and nuclear polymorphisms has provided insights into the long-discussed origins and relationships of olive cultivars. Like other agricultural crops, olives are currently experiencing a selection process driven by human activity. During the initial selection from wild to cultivated plants, genetic variability was likely enhanced due to the presence of a significant number of wild types throughout the Mediterranean Basin and beyond [32]. The wild relatives of olive cultivars could enhance the genetic foundation of cultivated olives. Certain agronomic traits, such as resistance to diseases like Verticillium dahliae Kleb and pests such as *Bactrocera oleae* Gmel, as well as low plant vigor and adaptability to challenging environments, are often absent in cultivated olives but can be observed in the wild germplasm [18, 51, 55].

In total, 50,684 protein-coding genes related to oleasters were predicted in the current genome assembly, with 47,124 of those genes (93%) confirmed through RNA sequencing (RNA-seq) data. Additionally, 31,245 genes were positioned on anchored pseudochromosomes. Around 90 million small RNA (sRNA) reads from six distinct tissues were utilized for annotating noncoding RNA (ncRNA). This analysis led to the identification of 498 conserved miRNA families and 125 novel miRNAs [18].

2.2 Morphology

The *Olea europaea*, whether in shrub or tree form, can grow to heights ranging from 3 to 15 meters. Its leaves are evergreen, leathery, entire, and covered with dense peltate scales. They can be narrowly elliptic, elliptic, ovate-oblong, or very narrowly elliptic, measuring between 3 to 9 cm in length and 0.3 to 3 cm in width. The inflorescences are axillary and can be cymose paniculate or racemose decussate. The calyx is small and divided into four broadly triangular lobes. The corolla features a short tube with four lobes that are valvate. There are two stamens with ellipsoid, exserted anthers. The ovary is either bottle-shaped or conoid. The drupes range from 0.5 to 4 cm in length, with a fleshy mesocarp that can be either thin or thick and a hard endocarp.

The *Olea europaea var. sylvestris* is known as the "wild olive of the Mediterranean area." It is characterized by its shrubby, twiggy, and spiny growth habit, along with smaller leaves that are ovate-oblong to elliptic and tiny fruits measuring less than 1 cm long, with a fleshy but thin mesocarp [46, 48].

Olive species produce small white flowers and small purple fruits (**Figure 10**). The flowers of the wild olive are hermaphroditic, containing both androecium and gynoecium. Many olive trees are andromonoecious, meaning that individual trees can produce both hermaphroditic and staminate flowers [56].

Figure 10.
Olive flowers.

Occasionally, individuals with both hermaphroditic and unisexual flowers are observed; however, staminate flowers (male flowers) are generally more prevalent than carpellate ones (female flowers). In cultivated varieties, andromonoecy is significantly more common compared to wild forms. *Olea europaea* is primarily recognized as a wind-pollinated species, with biotic pollinators likely playing a meaningful role only when other floral sources are scarce. When fruiting occurs via cross-pollination, numerous cultivars emerge, each exhibiting distinct sizes, shapes, and unique chemical and sensory properties [48, 57].

Morphological classification of olives is extensively utilized in research for describing fruit characteristics and distinguishing between cultivars. The International Olive Council Standards (IOC) categorize olive weight as low (< 2 g), medium (2–4 g), high (4–6 g), and very high (> 6 g), while fruit shape is assessed based on the aspect ratio of width to length, categorized as spherical (< 1.25), ovoid (1.25–1.45), or elongated (> 1.45). Olive fruit is a drupe that typically has a round to ovoid shape and measures 15–30 mm in diameter, changing from green in summer to light red in autumn. The olive drupe consists of the pericarp (mesocarp or flesh) and the endocarp (stone), with the flesh containing 96 to 98% of the total oil content, while the remaining oil is found within the seed [57].

3. Olive varieties for table and olive oil

G. Bartolini states that there are over 2600 olive cultivars globally, with 250 categorized as "commercial cultivars" by the International Olive Oil Council. S. Duran reports more than 1200 cultivars, while C. Breton notes over 2000. The FAO Plant Production and Protection Division estimates that the global olive germplasm collection comprises more than 1200 distinct cultivars along with over 3600 synonyms, reflecting a wealth of local cultivars and ecotypes [58, 59].

When selecting the variety mix for an olive grove, several factors must be considered: local agronomic practices and traditional conditions, climate, desired fruit or oil quality, productivity, and the varieties' tolerances or sensitivities to various pests and diseases.

Key characteristics of cultivars to evaluate include:

i. fruit and oil production yield.

ii. tolerance to abiotic factors (such as frost, drought, and salinity) or resistance to biotic factors (pathogens including Verticillium, peacock eye, olive knot, and Xylella and insects like olive flies and black scales) endemic to the area in question.

iii. Compatibility for self-pollination and inter-pollination. Thus, it's crucial to select at least two intercompatible cultivars within the same plot. It is generally recommended to plant three or four cultivars to mitigate the risk of one cultivar not flowering, which would result in insufficient fertile pollen for the others, and to account for potential variations in flowering times among different cultivars.

iv. Oil characteristics, specifically fatty acid composition. In certain cultivation environments where high temperatures persist during fruit ripening, some varieties

DOI: http://dx.doi.org/10.5772/intechopen.1008263

may produce oil with low oleic acid content (<70%). Therefore, careful selection of cultivars that can maintain a balanced unsaturated/saturated fatty acid ratio is vital, along with ensuring high levels of antioxidants (especially phenolic compounds and tocopherols) and certainly a desirable sensory profile.

v. Firmness and pigmentation of the fruit pulp. Olives with tougher pulp and limited or delayed pigmentation generally demonstrate greater resilience to mechanical damage during harvesting, transportation, or storage, which can otherwise lead to oil quality deterioration, i.e., increased acidity and oxidation [60].

3.1 Common olive varieties and main characteristics

Over 100 olive germplasm collections exist at various international, national, and regional levels, primarily in Mediterranean countries, aimed at conservation and breeding efforts. The olive tree displays numerous varieties that demonstrate significant or subtle phenotypic and genetic variations. These variations regarding size, color, oil content, fatty acid composition, and other characteristics have been well documented in the primary olive-producing nations (**Table 2** and **Figure 11**) [49, 53, 61–69].

Country	Variety	Notes
Spain	Picual	Green, distinctive pungency, sweet
	Arbequina	Small-sized, brown, productive variety that does not tolerate continental and dry climates well
	Arbosana	High yield, very suitable for dense planting
	Hojiblanca	Bitter oil
	Manzanilla	Small-sized, purplish-green, oval-shaped, low yield
	Cornicabra	Late-maturing variety, sensitive to tuberculosis and dacus
	Verdial	Good resistant
	Gordal	Thick fruit of often mediocre quality
	Negral	Good resistance to late frosts
Italy	Frantoio	Persistent fruitiness, low yield
	Leccino	Mild sweet oil, low yield
	Coratina	Low yield
	Nocellara del Belice/ Castelveltrano	Large fruit, buttery flavor
	Pendolino	Medium-sized, slightly fruity oil, universal pollinator
	Moraiolo	Medium-sized, high, and constant productivity; tolerance to drought and sea winds; susceptible to cold; exceptional organoleptic oil; sensitive to repilo and tuberculosis
	Carolea	Large, dual suitability both for oil and for table, both green and black, high rooting capacity, very tolerant of low temperatures, and high oil content
	Ascolana Tenera	High size and high pulp/stone ratio, table variety, cold-tolerant
	Belle de Espagne	Self-sterile variety

Country	Variety	Notes
Greece	Koroneiki	Small, mottled fruit, high yield, high quality, suitable for dense planting
	Kalamata	Large, black
	Thassos	Small, green, ripens on the tree
	Olympia	Very bitter oil
	Konservolia	High size and high pulp/stone ratio, high productivity but alternate bearing, quality of its fruits and adaptation, resistance to cold
Portugal	Galega	Medium-sized, considered productive but very alternate, tolerance to drought, sensitive to cold, salinity, and limestone, low oil content
Morocco	Picholine Marocaine	Medium-sized, dual-purpose variety, used both for oil and for table, both green and black, medium oil content, high oleic acid content, high resistance to freezing
France	Picholine	Green, medium-sized, slight hazelnut flavor in oil
	Lucques	Excellent table variety
	Salonenque	Slightly sensitive to dacus and cycloconium
	Tanche	Self-sterile variety
Türkiye	Ayvalık/Edremit	Purple, high oil content, durable, distinctive aroma, pollinator
	Gemlik	Medium-sized, productive, somewhat cold-resistant, and black in color, they are popular for olive oil production and table use
	Memecik	Durable, high oil content, fruity aroma. Memecik olives are generally preferred in olive oil production
	Erkence Hurma	Ripens on the tree
	Delice (wild)	Wild, tiny, won gold at 2016 NYOOC
	Domat	High size and high pulp/stone ratio, low oil content. Resistance to cold. It is especially preferred for table use. It is susceptible to Bactrocera oleae due to its soft flesh
	Sarıulak	Medium-sized, dual suitability both for oil and for table both green and black, alternation, yield is moderate, sensitive to cold, high oil content
	Topak Ulak	It is used both as a green table olive and for oil production. The olives are large, with small pits, and they have an attractive appearance. They can be very large olives, and the table quality of the fruits is very good
	Halhalı	It is generally used for oil production.
	Uslu	The fruit is long and oval in shape. The pit is slender and has a gordal appearance. The fleshy part is somewhat soft
Malta	Bidni	Vibrant purple color
Israel	Barnea	High yield, pest-resistant, green leaf taste
Tunisia	Chemlali	Drought-resistant, cold-tolerant, self-pollinating
	Chetoui	Medium-sized, resistant to cold and salinity, medium oil content, high polyphenol content
	Meski	High size and high pulp/stone ratio, reduced vigor and low rooting capacity, ease of separation from the stone
	Ouslati	Resistant to tuberculosis, the fruit comes off easily

Country	Variety	Notes
Syria	Sorani	Resistant to cold, drought, and salinity, with low resistance to detachment, both table olives and olive oil's oil content is high and of excellent quality
	Zaity	Resistant to cold and salinity, this variety is highly appreciated for its high oil yield (around 30%) band good quality oil
	Marruecos	Resistant to cold, drought, and salinity, both olive oil and table olive (green and black), high oleic acid content
USA	Ascola	Cold-resistant
	Mission	Large black, introduced to the continent by Spanish missionaries
Algeria	Chemlal	Sensitive to tuberculosis, late flowering, alternating
	Sigoise	Grown on a free-standing plant, good response to cutting, self-fertile
Argentina	Arauco	It is one of the leading varieties in table olive consumption. It was brought to Argentina from Spain and is also known as Criola. It has a somewhat elongated fruit with a pointed tip and a slightly curved pit
Chile	Azapa	It takes its name from the Azapa Valley in the Tarapacá region. Ninety percent of the olive-growing areas in Chile are located here. Azapa is a high-quality table olive with meaty, plump flesh and a thin skin
	Huasco	Another name for this olive is Sevillano de Huesco, and its flesh/pit ratio is 5/1
Australia	Verdial	It is generally considered green
Cyprus	Ada Yerlisi	Medium-sized, dark green in color, and oval in shape. The flesh of the fruit is juicy and soft. The taste of this variety is sweet and aromatic

Table 2.
Common olive varieties by country and their main characteristics.

Figure 11.
Gemlik (on the left side) and Topak Ulak (on the right side) varieties of Türkiye.

3.2 Primary olive varieties for olive oil production in the world

Most commercial olive cultivars for oil production have its source from Mediterranean countries. Some primary world cultivars used for olive oil production are listed in **Table 3** [53, 63, 67, 69].

Variety	Oil, %	Cold resistance	Fruit size	Weight of the fruits, g	Polyphenol content	Origin
Aglandau	23–27	Resistant	Medium	—	Medium	France
Arbequina	22–27	Resistant	Small	1–2	Low	Spain
Ascolona Tenera	13	Resistant	Large	10	—	Italy
Ayvalık	24–25	Medium Rezistant	Medium	4.5	High	Türkiye
Barnea	18	Sensitive	Medium	—	Medium	Israel
Belle de Espagne	—	Medium Resistant	Large	10–12	—	Italy
Bouteillan	20–25	Resistant	Medium	—	Medium	France
Chemlal	14–16	—	Medium	2.5	—	Algeria
Chemlali	26–28	—	Very Small	—	High	Tunisia
Chetoui	—	Resistant	Medium	3–4	High	Tunisia
Coratina	23–27	Resistant	Medium	4	Very High	Italy
Coratine	28–29	—	Medium	3.5–4	—	Italy
Cornicabra	23–27	Resistant	Medium	3–3.5	Very High	Spain
Farga	23–27	Resistant	Medium		Medium	Spain
Frantoio	23–26	Sensitive	Medium	2.5	Medium-High	Italy
Gemlik	22–29	Medium Resistant	Medium	2–3	High	Türkiye
Gordal	14–18	Medium Resistant	Large	11–12	—	Spain
Hojiblanca	23–28	Resistant	Medium	2–4	—	Spain
Koroneiki	24–28	Sensitive	Very Small	1.1	Very High	Greece
Leccino	22–27	Resistant	Medium	2.5	Medium	Italy
Lucquies	18–20	Resistant	Medium	4–5	—	France
Manzanilla	20	Sensitive	Medium	3.5–5	—	Spain
Mastoidis (Tsounati)	28–35	Resistant	Very Small	—	Very High	Greece
Maurino	20–25	Resistant	Medium	—	High	Italy
Memecik	24–25	Resistant	Large	5.2	High	Türkiye
Morailo	26–28	Sensitive	Medium	2.5	—	Italy
Negral	23–30	—	Medium	2.5–3	—	Spain
Ouslati	22–27	—	—	—	—	Tunisia
Pendolino	20–25	Resistant	Medium	2.5	Medium	Italy
Picholine	22–25	Moderate	Medium	3–4	High	France
Picholine Marroqui	19–25	—	Medium	3.5–5	—	Morocco
Picual	24–27	Resistant	Medium	2.5–3.5	Very High	Spain
Picudo	22–24	Resistant	Large	—	Low	Spain
Salonenque	20–22	Resistant	—	—	—	France

Variety	Oil, %	Cold resistance	Fruit size	Weight of the fruits, g	Polyphenol content	Origin
Sarıulak	18–19	Sensitive	Medium	3.76	Medium	Türkiye
Sigoise	—	—	Medium	3–3.5	—	Algeria
Taggiasca	22–27	Sensitive	Medium	—	Low	Italy
Tanche	25–30	Resistant	Medium	5–6	—	France
Verdial	21–31	Resistant	Medium	1.5–4.5	—	Spain

Table 3.
Primary oil-olive varieties in the world.

4. Nutritional and health value of olive fruit and olive oil

The Mediterranean diet is defined by a significant intake of unprocessed foods of plant origin, including fruits, vegetables, grains, and nuts, along with fish and poultry as the primary protein sources. It features low levels of meat, dairy products, eggs, animal fats, and discretionary foods. Extra virgin olive oil (EVOO) is appreciated as the primary fat source. Traditionally, the health benefits of EVOO have been linked to its high content of monounsaturated fatty acids (MUFA), particularly oleic acid, which can constitute up to 80% of its overall lipid profile [70, 71].

4.1 Chemical composition of olive and virgin olive oil

The main components of olives are water (60–75%), lipids (10–25%), reducing sugars (2–5%), and phenolic substances (1–3%) (**Figure 12**). Among the various components of olives, polyphenols stand out for their abundance and significance in both sensory qualities and potential health benefits [11]. The polyphenols found in olives can be categorized into five distinct classes [12, 13]:

1. *Acids*: Includes caffeic, gallic, and syringic acids.

2. *Alcohols*: Comprises tyrosol and hydroxytyrosol.

3. *Flavonoids*: Features luteolin-7-glucoside and cyanidin-3-glucoside.

4. *Secoiridoids*: Contains compounds such as oleuropein, which decreases during maturation, as well as demethyloleuropein and the dialdehydic form of elenolic acid associated with tyrosol and hydroxytyrosol—both of which increase as the fruit matures.

5. *Lignans*: Includes 1-acetoxypinoresinol and pinoresinol [72–74].

Additionally, olive leaves are commonly used to make herbal tea, primarily due to their high amount of phenolic compounds like oleuropein and hydroxytyrosol, which provide nutritional and medicinal benefits [18].

Epidemiological studies consistently indicate that higher olive oil consumption correlates with a decreased risk of various chronic illnesses. Research' highlights are the extensive health advantages of olive oil, particularly concerning cardiovascular

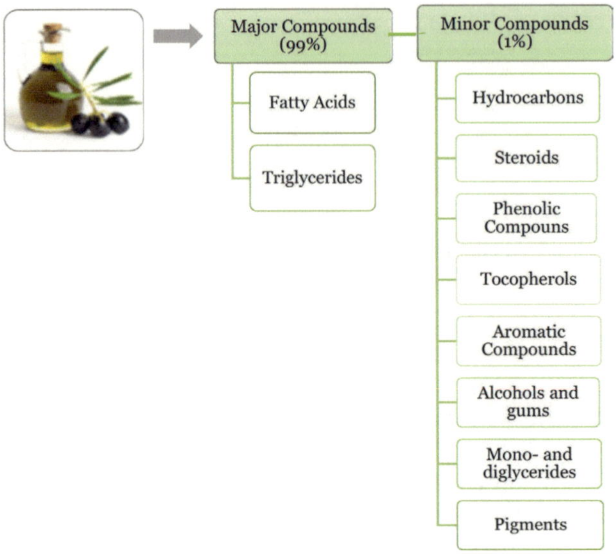

Figure 12.
Main compounds of olive and virgin olive oil [72].

disease, cancer, type 2 diabetes, body composition, blood pressure, inflammation, endothelial function, and hemostasis [75]. Key components such as oleic acid, tocopherols, and polyphenols are crucial for explaining olive oil's protective effects against diseases. Most health benefits associated with olive oil are linked to its minor components, especially oleuropein (OLE), hydroxytyrosol (HT), and oleocanthal (OLC), which are known for their antioxidant and anti-inflammatory properties, contributing to neuroprotection [76].

4.2 Table olives

"Table olives are produced from high-quality fruits of cultivated olive tree varieties selected for their attributes, including size, shape, flesh-to-stone ratio, palatability, firmness, and ease of separation from the pit, making them ideal for processing. These olives are treated to eliminate bitterness and can be preserved through natural fermentation or heat treatment, sometimes with the addition of preservatives, and are packaged either with or without brine" [77].

Oleuropein, the primary phenolic compound found in olive varieties, is largely responsible for the characteristic bitterness of olives. Notably, the concentration of oleuropein significantly decreases as olives ripen and undergo processing. To be consumable, olives must first undergo a debittering process, which involves the removal or breakdown of oleuropein using lye, certain microorganisms, or enzymes. In natural processing methods, olives are immersed in brine without any preliminary lye treatment, allowing bitterness to decrease during storage. Subsequently, fermentation occurs, imparting the olives with their unique texture and flavor. The use of lactic acid bacteria (LAB) strains has been suggested as a biological method for debittering olives, alongside the direct oxidation of oleuropein [78].

An important study conducted by Romero et al. [79] examined the triterpenic acids from seventeen different olive cultivars processed in Spanish, Greek, and Californian styles, revealing their anti-cancer properties. The study demonstrated

DOI: http://dx.doi.org/10.5772/intechopen.1008263

that olives placed directly in brine contained significantly higher levels of triterpenic acids compared to those treated with alkaline, which is specific to Spanish style. Furthermore, it was observed that naturally fermented black olives possess a much greater concentration of these bioactive compounds than olive oil, leading to the conclusion that these fermented olives deserve a reevaluation of their nutritional value [73].

4.3 "Liquid gold" olive oil

Olive oil is fundamentally fruit juice extracted from olives. Natural olive oil refers to "oils derived exclusively from the olives of the olive tree (*Olea europaea* L.) through mechanical or physical methods that do not alter the oil, particularly concerning thermal conditions." This oil is processed only through washing, decantation, centrifugation, and filtration, as defined by the International Olive Council (IOC). At room temperature, olive oil is clear, ranges from green to yellow in color, and can be enjoyed raw, offering an authentic aroma. It is highly valued not only for its taste but also for its health benefits, attributed to its content of unsaturated fatty acids, phenolic compounds, phytosterols, and carotenoids [80]. The yield and composition of olive oil are influenced by factors such as the growing conditions, olive variety, and fruit ripeness [81]. Olive oil is categorized into three primary groups based on factors like color, aroma, production method, and freshness suitable for human consumption [82]:

1. "*Extra virgin olive oil*: This is virgin olive oil with a free acidity level (expressed as oleic acid) not exceeding 0.80 grams per 100 grams, along with other physical and sensory characteristics that meet the established standards for this category.

2. *Virgin olive oil*: This type has a free acidity level not exceeding 2.0 grams per 100 grams and conforms to the relevant physical and sensory standards.

3. *Ordinary virgin olive oil*: This category encompasses virgin olive oil with a free acidity level of no more than 3.3 grams per 100 grams, meeting the corresponding standards."

Some virgin olive oils require processing before they can be consumed:

1. "*Lampante virgin olive oil*: This oil has a free acidity greater than 3.3 grams per 100 grams and/or does not meet the defined physical and sensory characteristics, making it suitable only for refining or technical uses.

2. *Refined olive oil*: This oil is produced from virgin olive oils through refining processes that preserve the initial glyceridic structure, with a free acidity of no more than 0.30 grams per 100 grams, conforming to the set standards.

3. *Riviera olive oil*: A blend of refined olive oil and virgin olive oils that is safe for consumption, with a free acidity not exceeding 1.00 gram per 100 grams, adhering to the specified characteristics."

Unlike refined seed and nut oils, which undergo processes to eliminate undesirable non-oil substances post-extraction, virgin olive oils can be consumed directly after being mechanically separated from the fruit. It is important to note that the refining process can lead to the loss of valuable aromatic compounds, micronutrients, and antioxidants, including polyphenols, tocopherols, sterols, and carotenoids [83]. Currently, many dietary guidelines advocate for the inclusion of healthy plant oils, such as olive or canola oil, in a balanced diet, with a preference for virgin olive oil over refined options [84].

4.4 A forgotten treasure: Making oil from wild olives

For an extended period, wild olives were regarded as having minimal agronomic significance and were rarely included in olive breeding initiatives, which primarily focused on intra-specific crosses among cultivars. Recently, however, the discovery of oleasters thriving in arid regions across various altitudes and soil types—resilient to diverse adverse environmental conditions—has highlighted the importance of these plants. Additionally, the recognition of the exceptional quality of extra virgin olive oil derived from wild trees has emphasized the need to conserve and effectively manage oleaster genetic resources. The conservation and safeguarding of wild olives have become increasingly crucial, especially with the growing loss of extensive old oleaster forests caused by deforestation, which is notably pervasive in certain Mediterranean regions [85].

In Spanish, feral olive trees are called *acebuches* (*Olea europaea var. sylvestris*), and their fruits are known as "*acebuchinas*.". The wild olive stands out for its ability to adapt to poor soils, becoming a valuable tool in reforestation and conservation programs. These wild olive trees are capable of capturing carbon dioxide from the atmosphere and preventing soil erosion, making them trees of great help in preserving the environment. *Acebuchinas* are small olives, between 1 and 3.5 cm long. When it grows, the fruit is green. During the year of maturation, it acquires a very characteristic black-purple color (**Figure 13**). An average of 4–6 kg of olives are needed to produce a liter of oil from commercial varieties, whereas for wild olive trees this amount increases to 15–20 kg. "Oil from wild olive trees has a particular flavor, a different taste. When you take it to a tasting panel, professional tasters do not know how to describe it" [86].

Figure 13.
Wild olives ©Pablo Esparza.

4.5 Olive and olive oil production

Due to the increasing demand in recent years for olive products like olive oil and table olives in the world, olive production is carried out economically in Mediterranean countries and other countries such as Argentina, Chile, and Peru, which have Mediterranean-like climates as well. According to 2023 FAO data, world olive grove areas increased by 5% between 2012 and 2021, reaching 10.34 million hectares. When the olive grove areas of the countries are examined, Spain with 25.4%, Tunisia with 12.4%, Italy with 10.9%, Morocco with 10.7%, Turkey with 8.6%, and Greece with 7.9% share the first ranks. World olive production tends to increase, although it fluctuates year to year. According to IOC 2024 data, world olive production reached 3,014,000 tons (**Figure 14**). Türkiye is in first place this year in olive production, having harvested 605,000 tons. Türkiye is followed by Spain with 414,000 tons of production, Greece with 325,000 tons, and Algeria with 256,000 tons [87–89].

World olive oil production is below average in 2023–2024 with 2.511 million tons (EC-June 2023), whereas 3.398 million tons in 2021–2022 (IOC-2023). While EU countries production remains below average, production in non-EU countries decreases by 14%. And consumer prices of olive oil are up 52% in one year. Drought and extreme weather conditions in the EU's south in 2023 significantly impacted olive yields [90]. During the 2022/2023 harvest season, Spain was the top producer with 665.8 thousand tons, followed by Türkiye with 380 thousand tons, Greece with 345 thousand tons, and Italy with 288.9 thousand tons. Collectively, these countries represent a substantial share of the world's olive oil production (**Figure 15**).

During the 2022–2023 harvest season, Spain was the top producer with 665.8 thousand tons, followed by Türkiye with 380 thousand tons, Greece with 345 thousand tons, and Italy with 288.9 thousand tons. Collectively, these countries represent a substantial share of the world's olive oil production [87].

World table olive production reached 3 million tons, 5 year-average, with a total of 3.02 million tons in this period. World table olive exports increased by 4% in 2022/23. The top importing countries were the USA, Brazil, and Canada. The increase in table olive prices in the world continues [88].

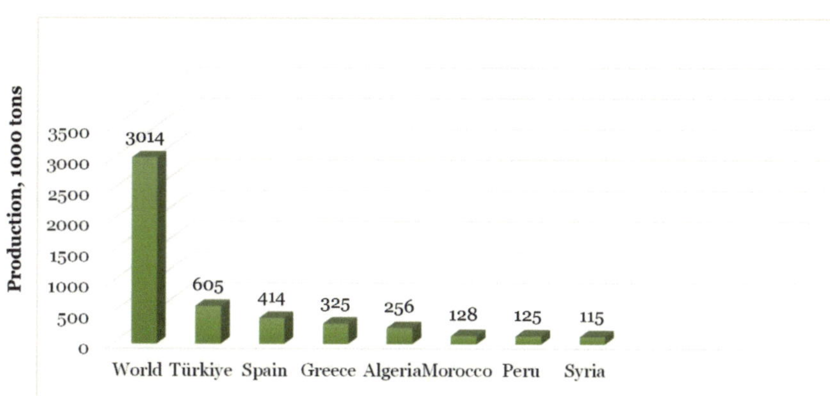

Figure 14.
Olive production in 2023–2024 [87].

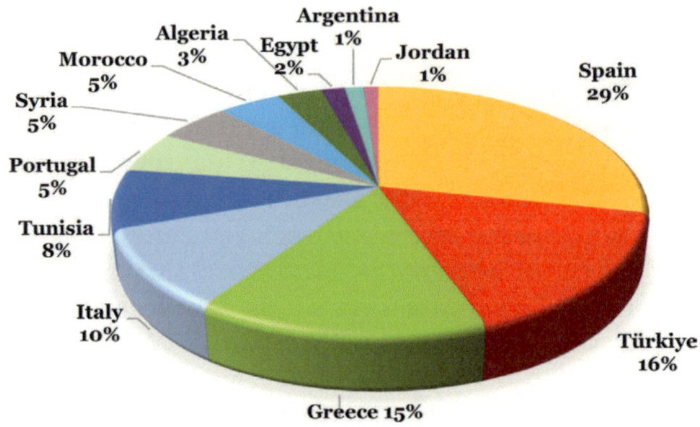

Figure 15.
World olive oil production in 2022–2023 [87].

5. Environmental factors for olive cultivation

Worldwide, olive tree cultivation is primarily restricted to the latitudinal range of 30–45°. This geographical zone indicates that climate plays a crucial role in the growth and development of olive trees. The ideal climate for olive cultivation resembles the Mediterranean climate, which acts as a bridge between the arid conditions of North Africa and the temperate, rainy climate of Central Europe. Generally, olive trees cannot endure temperatures below −8°C for more than a week. Additionally, excessively high summer temperatures can hinder their yield, with maximum temperatures exceeding ~30°C limiting production and photosynthesis when surpassing 40°C. A detailed climatological study of the Mediterranean Basin has shown that olive-growing regions are currently affected by temperatures during the coldest (January) and warmest (July) months. The optimal average monthly temperatures for olive cultivation are approximately 7°C in January and 25°C in July [91].

Precipitation is another vital climate factor. Around 90% of olive trees in the Mediterranean Basin rely mainly on rain-fed conditions. While olive trees can withstand drought, their presence in arid areas is restricted by annual rainfall below 350 mm, and water availability remains crucial for enhancing final yields. The minimum water requirement for olive cultivation is around 200 mm per year [92]. Consequently, olive growers adopt management strategies, such as sparse planting and rigorous pruning, to mitigate severe water stress. This underscores the significant role of rainfall in the economic sustainability of this crop, particularly given the typically dry summers in these regions. Therefore, farmers heavily rely on efficiently utilizing winter and spring rainfall for productive orchards. Soil characteristics, particularly soil water retention capacity, also significantly affect olive tree growth. Although olive trees adapt well to low-fertility, shallow, and poor soils, optimal conditions include deep, fertile soils with moderate moisture levels.

Other atmospheric elements, such as solar radiation, relative humidity, and wind, also affect olive orchard productivity. For instance, areas prone to strong winds should be avoided for olive cultivation, as cold, damp winds in spring can hinder flower fertilization and fruit development. Similarly, hot winds during summer can cause fruit drops, while dry winds may lead to early maturation and fruit shriveling. Since wind plays a crucial role in pollination, hot, dry winds can damage pollen grains [93].

Climate change is projected to lead to a notable decrease in suitable areas for olive cultivation and overall productivity, largely due to the strong adaptation of popular varieties to specific climatic conditions. Over centuries, olive growers have selected the most suitable varieties for their respective locations and climates. With anticipated climate change, growers may need to replace vulnerable varieties with those more resilient to changing conditions. The early identification of stress-tolerant varieties through rapid screening could significantly enhance agricultural sustainability and food security. It is essential to promote the cultivation of traditional varieties in potentially suitable regions to prevent replacement with less environmentally compatible alternatives. By increasing awareness of the valuable oleasters such as *Olea europaea sylvester* and *Olea europaea cuspidata*, these vital and irreplaceable resources for the future of olive species must be safeguarded [39, 91, 94]. Possible measures to improve crop resilience comprise selection of more resistant varieties, changing the management systems, especially irrigation, and analyzing predicted suitable areas or relocating plantations to higher altitudes [95].

6. Conclusions and future challenges

This chapter highlights the cultural, nutritional, and economic importance of olive varieties cultivated across the Mediterranean Basin and beyond. Olive diversity enriches both traditional and modern cuisines with unique flavors and nutritional value, serving as a cornerstone of the Mediterranean diet and providing numerous health benefits. Research emphasizes the adaptability of olive cultivars to various climate and environmental stressors, supporting the sustainability of olive production despite the challenges posed by global climate change.

However, several challenges lie ahead. As the impacts of climate change intensify, more frequent droughts, extreme temperatures, and shifts in agricultural zones are expected. Future research should prioritize the development of olive varieties with greater resistance to extreme weather conditions and diverse soil types. Preserving genetic diversity will be crucial for breeding resilient cultivars and ensuring the sustainability of olive production in the long term.

Promoting sustainable agricultural practices will be another key challenge. As olive growers face increasing environmental and economic pressures, advances in biotechnology will play a pivotal role. Innovative solutions, such as Internet of Things (IoT)-Agro: Smart Farming Systems, genetic tools, and eco-friendly farming techniques, will support the sector's adaptation to evolving conditions. Addressing these challenges holistically will secure the future of olive cultivation while preserving the cultural heritage and economic value of this ancient crop.

Author details

Perihan Adun and Şebnem Güler*
Faculty of Agriculture, Department of Food Engineering, Near East University, Nicosia, North Cyprus

*Address all correspondence to: sebnem.guler@neu.edu.tr

References

[1] Book of Genesis. Available from: https://www.vatican.va/archive/bible/genesis/documents/bible_genesis_en.html

[2] Zohary D, Hopf M, Weiss E. Domestication of Plants in the Old World: The Origin and Spread of Domesticated Plants in Southwest Asia, Europe, and the Mediterranean Basin. 4th ed. Oxford, United Kingdom: Oxford University Press; 2012. DOI: 10.1093/acprof:osobl/9780199549061.001.0001

[3] Kaniewski D, Van Campo E, Boiy T, Terral JF, Khadari B, Besnard G. Primary domestication and early uses of the emblematic olive tree: Palaeobotanical, historical and molecular evidence from the Middle East. Biological Reviews. 2012;**87**(4):885-899. DOI: 10.1111/j.1469-185X.2012.00229.x

[4] Besnard G, Khadari B, Navascués M, Fernández-Mazuecos M, El Bakkali A, Arrigo N, et al. The complex history of the olive tree: From late quaternary diversification of Mediterranean lineages to primary domestication in the northern Levant. Proceedings of the Royal Society B: Biological Sciences. 2013;**280**(1756):20122833. DOI: 10.1098/rspb.2012.2833

[5] Grigg D. Olive oil, the Mediterranean and the world. GeoJournal. 2001;**53**:163-172

[6] Clodoveo ML, Camposeo S, De Gennaro B, Pascuzzi S, Roselli L. In the ancient world, virgin olive oil was called "liquid gold" by Homer and "the great healer" by Hippocrates. Why has this mythic image been forgotten? Food Research International. 2014;**62**:1062-1068. DOI: 10.1016/j.foodres.2014.05.034

[7] World History Encyclopedia. Available from: https://www.worldhistory.org/

[8] Loumou A, Giourga C. Olive groves: "The life and identity of the Mediterranean". Agriculture and Human Values. 2003;**20**:87-95

[9] Routes of the Olive Tree. Available from: https://www.coe.int/en/web/cultural-routes/the-routes-of-the-olive-tree [Accessed: September 16, 2024]

[10] Letendre D. Les mots de l'insoumis. Liberte. 2014;**47**:278

[11] Fraser J, Cartwright C, Discovering a 4,500-Year-Old Olive Oil Factory in Jordan. 2020. Available from: https://www.britishmuseum.org/blog/discovering-4500-year-old-olive-oil-factory-jordan [Accessed: September 17, 2024]

[12] Banning EB. So fair a house: Göbekli Tepe and the identification of temples in the pre-pottery Neolithic of the near east. Current Anthropology. 2011;**52**(5):619-660. DOI: 10.1086/661207

[13] Dietrich L, Meister J, Dietrich O, Notroff J, Kiep J, Heeb J, et al. Cereal processing at early Neolithic Göbekli Tepe, southeastern Turkey. PLoS One. 2019;**14**(5):1-34. DOI: 10.1371/journal.pone.0215214

[14] Knitter D, Braun R, Clare L, Nykamp M, Schütt B. Göbekli Tepe: A brief description of the environmental development in the surroundings of the UNESCO world heritage site. Land. 2019;**8**(4):72. DOI: 10.3390/land8040072

[15] Neef R. Overlooking the steppe-forest: A preliminary report on the

botanical remains from early Neolithic Göbekli Tepe (southeastern Turkey). Neo-Lithics. 2002;**3**:13-16

[16] Kennedy L, The Prehistoric Ages: How Humans Lived Before Written Records. 2024. Available from: https://www.history.com/news/prehistoric-ages-timeline [Accessed: September 9, 2024]

[17] Sarma J. The origin of food cultivation in the middle-east and the human environment dynamics. International Journal of History and Research (IJHR), 2018;**8**(2):11-18. ISSN (P): 2249-6963; ISSN (E): 2249-8079 © TJPRC Pvt. Ltd., India

[18] Breton CM, Warnock P, Jean Bervillé A, "Origin and history of the olive." Olive Germplasm. In: Muzzalupo I, editor. The Olive Cultivation, Table Olive and Olive Oil Industry in Italy. London, UK: IntechOpen; 2012. DOI: 10.5772/3314

[19] Julca I, Marcet-Houben M, Cruz F, Gómez-Garrido J, Gaut BS, Díez CM, et al. Genomic evidence for recurrent genetic admixture during the domestication of Mediterranean olive trees (*Olea europaea* L.). BMC Biology. 2020;**18**:1-25. DOI: 10.1186/s12915-020-00881-6

[20] Diez CM, Trujillo I, Martinez-Urdiroz N, Barranco D, Rallo L, Marfil P, et al. Olive domestication and diversification in the Mediterranean basin. New Phytologist. 2015;**206**(1):436-447. DOI: 10.1186/s12915-020-00881-6

[21] Julca I, Vargas P, Gabaldón T. Phylogenomics of the Olea europaea complex using 15 whole genomes supports recurrent genetic admixture together with differentiation into seven subspecies. BMC Biology. 2023;**21**(1):85. DOI: 10.1186/s12915-023-01583-5

[22] Besnard G, Terral JF, Cornille A. On the origins and domestication of the olive: A review and perspectives. Annals of Botany. 2018;**121**(3):385-403. DOI: 10.1093/aob/mcx145

[23] Vossen P. Olive oil: History, production, and characteristics of the world's classic oils. HortScience. 2007;**42**(5):1093-1100. DOI: 10.21273/HORTSCI.42.5.1093

[24] Di Giovacchino L. Technological aspects. In: Handbook of Olive Oil. Analysis and Properties. Aspen: Aspen Publications; 2000. pp. 17-59. DOI: 10.1007/978-1-4757-5371-4

[25] Uylaşer V, Yildiz G. The historical development and nutritional importance of olive and olive oil constituted an important part of the Mediterranean diet. Critical Reviews in Food Science and Nutrition. 2014;**54**(8):1092-1101. DOI: 10.1080/10408398.2011.626874

[26] Terral JF, Bonhomme V, Pagnoux C, Ivorra S, Newton C, Paradis L, et al. The shape diversity of olive stones resulting from domestication and diversification unveils traits of the oldest known 6500-years-old table olives from Hishuley Carmel site (Israel). Agronomy. 2021;**11**(11):2187. DOI: 10.3390/agronomy11112187

[27] Koh AJ, Betancourt PP. Wine and olive oil from an early Minoan I hilltop fort. Mediterranean Archaeology and Archaeometry. 2010;**10**(2):15-15

[28] Hamilakis Y. The Nation and its Ruins: Antiquity, Archaeology, and National Imagination in Greece. online edn. Oxford, UK: Oxford University Press; 2007. 360 p. DOI: 10.1093/oso/9780199230389.002.0001

[29] Riley FR. Olive oil production on bronze age crete: Nutritional properties,

processing methods and storage life of Minoan olive oil. Oxford Journal of Archaeology. 2002;**21**(1):63-75. DOI: 10.1111/1468-0092.00149

[30] Kapellakis IE, Tsagarakis KP, Crowther JC. Olive oil history, production and by-product management. Reviews in Environmental Science and Bio/Technology. 2008;**7**:1-26. DOI: 10.1007/s11157-007-9120-9

[31] Fiorino P, Nizzi Grifi F. Olive maturation and variations in certain oil constituents. Ed. Francaise. International Olive Oil Council. Madrid, Spain: Olivae, 1991;**35**:25-33

[32] Mariotti R, Belaj A, de la Rosa R, Muleo R, Cirilli M, Forgione I, et al. Genealogical tracing of *Olea europaea* species and pedigree relationships of var. europaea using chloroplast and nuclear markers. BMC Plant Biology. 2023;**23**:452. DOI: 10.1186/s12870-023-04440-3

[33] Chazau-Gillig S. The civilization of olive trees and cereals. Olivae. 1994;**53**:14

[34] Melena JL. Olive Oil and Other Sorts of Oil in the Mycenaean Tablets. Minos: Revista de Filología Egea; 1983. p. 18

[35] Davaras K. Guide to Cretan Antiquities. Park Ridge, NJ, USA: Noyes Press; 1976

[36] Psilakis N, Psilaki M, Kastanas I. To Elaiolado. (Olive oil). Karmanor, Greece: Greek Academy of Taste; 1999

[37] Kopaka K. A brief historical review of archaeological research on olive oil and wine in Crete in the bronze age. Ariadne. 2005;**11**:9-37. DOI: 10.26248/ariadne.v11i0.971

[38] Unver T, Wu Z, Sterck L, Turktas M, Lohaus R, Li Z, et al. Genome of wild olive and the evolution of oil biosynthesis. Proceedings of the National Academy of Sciences. 2017;**114**(44):E9413-E9422. DOI: 10.1073/pnas.1708621114

[39] Arenas-Castro S, Gonçalves JF, Moreno M, Villar R. Projected climate changes are expected to decrease the suitability and production of olive varieties in southern Spain. Science of the Total Environment. 2020;**709**:136161. DOI: 10.1016/j.scitotenv.2019.136161

[40] Rene JH, Garcia-Bertrand R, Ancestral DNA. Human Origins, and Migrations. 1st ed. Amsterdam (The Netherlands) and New York: Elsevier; 2018. 588 p. Paperback ISBN: 9780128041246, eBook ISBN: 9780128041284

[41] IPCC. In: Core Writing Team, Lee H, Romero J, editors. Climate Change 2023: Synthesis Report. Contribution of Working Groups I, II and III to the Sixth Assessment Report of the Intergovernmental Panel on Climate Change. Geneva, Switzerland: IPCC; 2023. pp. 35-115. DOI: 10.59327/IPCC/AR6-9789291691647

[42] Moriondo M, Stefanini FM, Bindi M. Reproduction of olive tree habitat suitability for global change impact assessment. Ecological Modelling. 2008;**218**(1-2):95-109. DOI: 10.1016/j.ecolmodel.2008.06.024

[43] Ponti L, Gutierrez AW, Ruti PM, Dell'aquila A. Fine-scale ecological and economic assessment of climate change on olive in the Mediterranean basin reveals winners and losers. Proceedings of the National Academy of Sciences. 2014;**111**(15):5598-5603. DOI: 10.1073/pnas.1314437111

[44] De Andreis P. Climate Change Taking a Toll on Andalusian Olive

Oil Production. 2024. Available from: https://www.oliveoiltimes.com/production/climate-change-andalusia-olive-oil-production [Accessed: August 12, 2024]

[45] International Olive Oil Council. World Olive Encyclopaedia. 1st English ed. Spain: International Olive Oil Council; 1996

[46] Green PS. A revision of *Olea L.* (Oleaceae). Kew Bulletin. 2002;**57**(1):91-140

[47] Tourvas N, Ganopoulos I, Koubouris G, Kostelenos G, Manthos I, Bazakos C, et al. Wild and cultivated olive tree genetic diversity in Greece: A diverse resource in danger of erosion. Frontiers in Genetics. 2023;**14**:1298565. DOI: 10.3389/fgene.2023.1298565

[48] Kassa A, Konrad H, Geburek T. Molecular diversity and gene flow within and among different subspecies of the wild olive (*Olea europaea* L.): A review. Flora. 2019;**250**:18-26. DOI: 10.1016/j.flora.2018.11.014

[49] Boskou D. Olive and Olive Oil Bioactive Constituents Book. 1st ed. Urbana, Illinois: AOCS Press; 2015. 422 p. eBook ISBN: 9781630670429

[50] Cruz F, Julca I, Gómez-Garrido J, Loska D, Marcet-Houben M, Cano E, et al. Genome sequence of the olive tree, *Olea europaea*. GigaScience. 2016;**5**:29

[51] Lumaret R, Ouazzani N, Michaud H, Vivier G, Deguilloux MF, Di Giusto F. Allozyme variation of oleaster populations (wild olive tree) (*Olea europaea* L.) in the Mediterranean basin. Heredity. 2004;**92**:343-351

[52] Baldoni L, Tosti N, Ricciolini C, Belaj A, Arcioni S, Pannelli G, et al. Genetic structure of wild and cultivated olives in the Central Mediterranean basin. Annals of Botany. 2006;**98**(5):935-942

[53] Rallo L, Barranco D, Caballero JM, Del Rio C, Martín A, Tous J, et al. Variedades de olivo en España. Sevilla: Junta de Andalucía, Consejería de Agricultura Pesca y Alimentación ; Madrid: Ministerio de Agricultura Pesca y Alimentación, Secretaría General Técnica, Centro de Publicaciones: Mundi Prensa; 2005. 478 p. ISBN: 84-8476-192-4

[54] Clark PU, Dyke AS, Shakun JD, Carlson AE, Clark J, Wohlfarth B, et al. The last glacial maximum. Science. 2009;**325**:710-714. DOI: 10.1126/science.1172873

[55] Belaj A, León L, Satovic Z, de la Rosa R. Variability of wild olives (*Olea europaea* subsp. europaea var. sylvestris) analyzed by agro-morphological traits and SSR markers. Scientia Horticulturae. 2011;**129**(4):561-569. DOI: 10.1016/j.scienta.2011.04.025

[56] Cuevas J, Vito SP. The role of staminate flowers in the breeding system of *Olea europaea* (Oleaceae): An andromonoecious, wind-pollinated taxon. Annals of Botany. 2004;**93**(5):547-553

[57] Silva BS, Schmiele M. From olive to olive oil: A general approach. Research, Society and Development. 2021;**10**(3):e32210313408. DOI: 10.33448/rsd-v10i3.13408

[58] Ebiad R, Abu-Quaoud H. Morphological and biological characterization of three olive "*Olea europaea L.*". Cultivars in Palestine. Jordan Journal of Agricultural Sciences. 2014;**10**(1):130-143

[59] Slobodova N, Sharko F, Gladysheva-Azgari M, Tsiupka V, Boulygina E,

Rastorguev S, et al. Genetic diversity of common olive (*Olea europaea* L.) cultivars from Nikita botanical gardens collection revealed using RAD-Seq method. Genes. 2023;**14**:1323. DOI: 10.3390/genes14071323

[60] Proietti P, Regni L, Climate Change Mitigation through a Sustainable Supply Chain for the Olive Oil Sector Handbook. Italy: EU; 2019. 199 p. Available from: www.olive4climate.eu

[61] Blatchly R, Zeytinlikten Sofraya Zeytinyağının Hikayesi (Çev. Delen Nircan, Z.). Istanbul, Türkiye: Türkiye İş Bankası Kültür Yayınları; 2020. ISBN: 9786257999335

[62] Barranco D, Rallo L. Olive cultivars in Spain. HortTechnology. 2000;**10**(1):107-110

[63] Kaya HB, Cetin O, Kaya HS, Sahin M, Sefer F, Tanyolac B. Association mapping in Turkish olive cultivars revealed significant markers related to some important agronomic traits. Biochemical Genetics. 2016;**54**:506-533. DOI: 10.1007/s10528-016-9738-9

[64] Aktan N, Yıldırım HK. Sofralık Zeytin Teknolojisi. İzmir: Meta Basım; 2012. 138 s

[65] Barranco ND, Fernández Escobar R, Rallo RL. Capitulo 3, Variedades y Patrones. In: El cultivo del olivo. 6th ed. Co-published by Ediciones Mundi-Prensa, Madrid, and Junta de Andalucía, Sevilla; 2008. 846 p.

[66] Santos-Antunes F, León L, de la Rosa R, Alvarado J, Mohedo A, Trujillo I, et al. The length of the juvenile period in olive as influenced by vigor of the seedlings and the precocity of the parents. HortScience. 2005;**40**(5):1213-1215

[67] Kiritsakis K, Evangelou E, Sakellaropoulos N, Kiritsakis A. Olive oil processing, categories, nutritional benefits, and byproducts. In: Siddiq M, Uebersax MA, editors. Handbook of Vegetables and Vegetable Processing. Oxford, OX4 2DQ, UK Wiley; 2018. p. 745-760. DOI: 10.1002/9781119098935

[68] Canözer Ö. Standart Zeytin Çeşitleri Kataloğu. Ankara, Türkiye: TC Tarım ve Köyişleri Bakanlığı, Mesleki Yayınlar Serisi, Genel; 1991. 334 p

[69] Kaya H, Sefer F, Mete N, Çetin Ö, Hakan M, Şahin M, et al. Türkiye Zeytin Çeşit Kataloğu. İzmir: Gıda Tarım ve Hayvancılık Bakanlığı, Zeytin Araştırma Enstitüsü Müdürlüğü, Bassaray Matbaası; 2015. ISBN: 978-605-9175-04-3

[70] Trichopoulou A. Mediterranean diet, traditional foods, and health: Evidence from the Greek EPIC cohort. Food and Nutrition Bulletin. 2007;**28**(2):236-240. DOI: 10.1177/156482650702800213

[71] Yubero-Serrano EM, Lopez-Moreno J, Gomez-Delgado F, Lopez-Miranda J. Extra virgin olive oil: More than a healthy fat. European Journal of Clinical Nutrition. 2019;**72**(Suppl 1):8-17. DOI: 10.1038/s41430-018-0304-x

[72] Göğüş F, Özkaya MT, Ötleş S. Zeytinyağı. Ankara: EFLATUN; 2009

[73] Conte P, Fadda C, Del Caro A, Urgeghe PP, Piga A. Table olives: An overview on effects of processing on nutritional and sensory quality. Food. 2020;**9**(4):514. DOI: 10.3390/foods9040514

[74] Jimenez-Lopez C, Carpena M, Lourenço-Lopes C, Gallardo-Gomez M, Lorenzo JM, Barba FJ, et al. Bioactive compounds and quality of extra virgin olive oil. Food. 2020;**9**(8):1014. DOI: 10.3390%2Ffoods9081014

[75] Chiavarini M, Rosignoli P, Giacchetta I, Fabian R. Health outcomes associated with olive oil intake: An umbrella review of meta-analyses. Food. 2024;**13**(16):2619. DOI: 10.3390%2Ffoods13162619

[76] Gonçalves M, Vale N, Silva P. Neuroprotective effects of olive oil: A comprehensive review of antioxidant properties. Antioxidants. 2024;**13**(7):762. DOI: 10.3390/antiox13070762

[77] IOC International Olive Oil Council (IOC). Trade Standard Applying to Table Olives, COI/OT/NC no. 1. Madrid – España: IOC; 2004.

[78] Değirmencioğlu N. Modern techniques in the production of table olives. In: Boskou D, Clodoveo ML, editors. Products from Olive Tree. London, UK: IntechOpen; 2016. p. 215. DOI: 10.5772/64988

[79] Romero C, García A, Medina E, Ruíz-Méndez MV, de Castro A, Brenes M. Triterpenic acids in table olives. Food Chemistry. 2010;**118**:670-674. DOI: 10.1016/j.foodchem.2009.05.037

[80] Karagözlü M, Adun P, Ozsoz M, Aşır S. Detection of adulteration in cold-pressed olive oil by voltammetric analysis of alpha-tocopherol on single-use pencil graphite electrode. International Journal of Food Science & Technology. 2024;**59**(3):1781-1788. DOI: 10.1111/ijfs.16931

[81] Jiaojiao LV, Jiang C, Wu W, Mao K, Wei Q, Zheng Y, et al. The gapless genome assembly and multi-omics analyses unveil a pivotal regulatory mechanism of oil biosynthesis in the olive tree. Horticulture Research. 2024;**11**:8. DOI: 10.1093/hr/uhae168

[82] IOC International Olive Oil Council (IOC). Trade Standard Applying to Olive Oils and Olive Pomace Oils, COI/T.15/NC No 3. Madrid- España: IOC International Olive Oil Council (IOC); 2019. 18p.

[83] SzydłOwska-Czerniak A. Rape seed and its products—Sources of bioactive compounds: A review of their characteristics and analysis. Critical Reviews in Food Science and Nutrition. 2013;**53**(4):307-330

[84] Donat-Vargas C, Lopez-Garcia E, Banegas JR, Martínez-González MÁ, Rodríguez-Artalejo F, Guallar-Castillón P. Only virgin type of olive oil consumption reduces the risk of mortality. Results from a Mediterranean population-based cohort. European Journal of Clinical Nutrition. 2023;**77**(2):226-234

[85] Fanelli V, Mascio I, Falek W, Miazzi MM, Montemurro C. Current status of biodiversity assessment and conservation of wild olive (*Olea europaea* L. subsp. *Europaea* var. sylvestris). Plants. 2022;**11**(4):480. DOI: 10.3390/plants11040480

[86] A Forgotten Treasure: Making Oil from Wild Olives. Available from: https://www.oliveoiltimes.com [Accessed: September 27, 2024]

[87] IOC. World Market of Olive Oil and Table Olives-Data from June 2024. Available from: IOOC@internationaloliveoil.org/world-market-of-olive-oil-and-table-olives-data-from-June-2024 [Accessed: September 12, 2024]

[88] AEPDI/TEPGE. Table Olives and Olive Oil Agricultural Products Market Reports. Agricultural Economics and Policy Development Institute: Ankara-Türkiye; 2024. Available from: https://arastirma.tarimorman.gov.tr/tepge

[89] EC, Market Situation in the Olive Oil and Table Olive Sectors, Export Group

for Agricultural Markets-Arable Crops and Olive Oil. 2024. Available from: https://agridata.ec.europa.eu/extensions/dataportal/home.html [Accessed: September 12, 2024]

[90] Nteve GM, Kostas S, Polidoros AN, Madesis P, Nianiou-Obeidat I. Adaptation mechanisms of olive tree under drought stress: The potential of modern omics approaches. Agriculture. 2024;**14**:579. DOI: 10.3390/agriculture14040579

[91] Fraga H, Moriondo M, Leolini L, Santos JA. Mediterranean olive orchards under climate change: A review of future impacts and adaptation strategies. Agronomy. 2021;**11**(1):56. DOI: 10.3390/agronomy11010056

[92] Alfieri SM, Riccardi M, Menenti M, Basile A, Bonfante A, De Lorenzi F. Adaptability of global olive cultivars to water availability under future Mediterranean climate. Mitigation and Adaptation Strategies for Global Change. 2017;**24**:435-466. DOI: 10.1007/s11027-018-9820-1

[93] Griggs WH, Hartmann HT, Bradley MV, Iwakiri BT, Whisler JE. Pollination in California. In: California Agricultural Experiment Station. Berkeley, CA, USA: University of California; 1975. p. 49

[94] Zunino L, Cubry P, Sarah G, Mournet P, El Bakkali A, Aqbouch L, et al. Genomic evidence of genuine wild versus admixed olive populations evolving in the same natural environments in western Mediterranean basin. PLoS One. 2024;**19**(1):1-18. DOI: 10.1371/journal.pone.0295043

[95] Honorio F, Aguirado C, Paniagua LL, García-Martín A, Rebollo L, Rebollo FJ. Exploring the climate and topography of olive orchards in extremadura, southwestern Spain. Land. 2024;**13**:495. DOI: 10.3390/land13040495